WHAT IT TAKES

TAKES

AIR FORCE COMMAND OF JOINT OPERATIONS

MICHAEL SPIRTAS

THOMAS-DURELL YOUNG

S. REBECCA ZIMMERMAN

The research described in this report was sponsored by the United States Air Force under Contract FA7014-06-C-0001. Further information may be obtained from the Strategic Planning Division, Directorate of Plans, Hq USAF.

Library of Congress Cataloging-in-Publication Data

Spirtas, Michael.
  What it takes : Air Force command of joint operations / Michael Spirtas, Thomas-Durell Young, S. Rebecca Zimmerman.
    p. cm.
  Includes bibliographical references.
  ISBN 978-0-8330-4614-7 (pbk. : alk. paper)
  1. Unified operations (Military science) 2. United States. Air Force. 3. Command of troops. I. Young, Thomas-Durell. II. Zimmerman, S. Rebecca. III. Title.

U260.S667 2009
355.3'30410973—dc22

2009001048

The RAND Corporation is a nonprofit research organization providing objective analysis and effective solutions that address the challenges facing the public and private sectors around the world. RAND's publications do not necessarily reflect the opinions of its research clients and sponsors.

RAND® is a registered trademark.

Cover design by Peter Soriano.

© Copyright 2009 RAND Corporation

All rights reserved. No part of this book may be reproduced in any form by any electronic or mechanical means (including photocopying, recording, or information storage and retrieval) without permission in writing from RAND.

Published 2009 by the RAND Corporation
1776 Main Street, P.O. Box 2138, Santa Monica, CA 90407-2138
1200 South Hayes Street, Arlington, VA 22202-5050
4570 Fifth Avenue, Suite 600, Pittsburgh, PA 15213-2665
RAND URL: http://www.rand.org/
To order RAND documents or to obtain additional information, contact
Distribution Services: Telephone: (310) 451-7002;
Fax: (310) 451-6915; Email: order@rand.org

Preface

When appropriate, the U.S. Air Force needs to be prepared to supply joint task force (JTF) headquarters. This monograph seeks to help Air Force personnel understand the requirements[1] of an effective JTF headquarters and to identify the broad outlines for how the Air Force can build and maintain this capability. It considers the nature of JTF command, surveys command-related developments in other services and in other elements of the defense community, and examines four JTF operations. It raises issues for the Air Force to consider and offers a set of recommendations aimed at enhancing the Air Force's ability to staff and run JTF headquarters.

The research documented here should be of interest to a wide group of Air Force personnel involved in the development and function of the service's command organizations, including component–Numbered Air Force (C-NAF) staff, those working on command policy, and more generally those interested in the role of air power in joint operations. It should also be of interest to other members of the defense community seeking to understand issues related to command and to the future of joint military operations.

The research reported here was sponsored by the Deputy Chief of Staff for Operations, Plans and Requirements, Headquarters, U.S. Air Force. The research was conducted within the Strategy and Doctrine Program of RAND Project AIR FORCE for a fiscal year 2007

---

[1] By the term "requirements" we do not mean to imply that we have derived formal Department of Defense requirements for JTF command. We use the term to refer to the necessary characteristics of a successful JTF command.

study "Joint Warfighting Headquarters." The principal research was completed in 2007 and builds on work done at the RAND Corporation on the issue of command. Previous RAND reports in this area include the following:

- *Enhancing Army Joint Force Headquarters Capabilities*, by Timothy Bonds, Myron Hura, and Thomas-Durell Young (MG-675-A, forthcoming). This monograph is aimed at helping the U.S. Army improve its ability to command and control joint, interagency, and multinational forces.
- *Learning Large Lessons: The Evolving Roles of Ground Power and Air Power in the Post–Cold War Era*, by David E. Johnson (MG-405-1-AF, 2007). Because joint doctrine frequently reflects a consensus view rather than a truly integrated joint perspective, the author recommends that joint doctrine—and the processes by which it is derived and promulgated—be overhauled.
- *Command Concepts: A Theory Derived from the Practice of Command and Control*, by Carl H. Builder, Steven C. Bankes, and Richard Nordin (MR-775-OSD, 1999). Through six historical case studies of modern battles, this book explores the implications of the theory for the professional development of commanders and for the design and evaluation of command and control architectures.

RAND Project AIR FORCE

RAND Project AIR FORCE (PAF), a division of the RAND Corporation, is the U.S. Air Force's federally funded research and development center for studies and analyses. PAF provides the Air Force with independent analyses of policy alternatives affecting the development, employment, combat readiness, and support of current and future aerospace forces. Research is conducted in four programs: Force Modernization and Employment; Manpower, Personnel, and Training; Resource Management; and Strategy and Doctrine.

Additional information about PAF is available on our Web site: http://www.rand.org/paf/

Contents

Preface iii
Figures ix
Tables xi
Summary xiii
Acknowledgments xix
Abbreviations xxiii

CHAPTER ONE
Introduction and Purpose 1

CHAPTER TWO
Background 3
JTFs in Theory 3
JTFs in Practice 6
The Problem 11
The Objective 16

CHAPTER THREE
Command Concepts 19
Themes 20
   Employment and Management 20
   Time and Function 22
Army 23
Navy 29
Marine Corps 33

Joint Force Command... 34
Air Force 36

CHAPTER FOUR
Lessons from Past JTFs... 41
JTF–Atlas Response: The Benefits of Preparation and Presence 42
JTF–Unified Assistance (CSF-536): Mixed Modes of Control.............. 47
JTF–Noble Anvil: The Questionable Joint Task Force 54
JTF–Southwest Asia: *Groundhog Day*..59
Summary 63

CHAPTER FIVE
Requirements 65
Build 65
    Develop Commanders... 65
    Build Staffs 66
Prepare.. 67
    Identify Missions.. 67
    Exercise.. 68
    Engage Partners 68
Execute 69
    Build and Maintain Partnerships 69
    Staff the Headquarters 69
    Issue Orders.. 70
    Gain and Maintain Situational Awareness.................................. 70
    Orchestrate Efforts... 71
    Assess and Adjust 72

CHAPTER SIX
Issues 73
Separate or Combine Employment and Management........................ 73
Organize Around Time or Function... 74
Determine When the Air Force Leads a JTF and How Many Types
    of JTF Headquarters Does It Need 75
Determine How Many JTF-Capable NAFs the Air Force Needs........... 76

Determine How the Air Force Would Simultaneously Provide
  C-NAF and JTF Headquarters..78
Determine How the Air Force Would Man the Bulk of JTF
  Headquarters Positions...79
Determine How the JTF Headquarters Would Incorporate
  Other Services and Non-DoD Partners....................................79

CHAPTER SEVEN
Recommendations..81
Systems..81
  Acquire Necessary Systems...81
  Determine the Desired Approach Toward Reach Back..................82
People..82
  Reward Officers' Deep Experience with Joint, Interagency, and
    International Partners..82
  Reorient Professional Military Education..................................83
  Assign Competitive Personnel to AFFOR Staffs........................ 84
  Train AFFOR Staffs... 86
Processes.. 87
  Designate JTF-Capable Organizations...................................... 87
  Use Exercise Programs... 87
  Place More Emphasis on Planning... 88
  Write a Directive on Air Force JTF Operations..........................89
  Learn JTF Headquarters Processes... 90
  Create a Capability to Deploy Headquarters............................. 90
  Create a Champion for Air Force Command............................. 91
Conclusion.. 92

APPENDIXES
A. **Joint Task Forces Since 1990**...93
B. **Joint Manning Document Data from Selected Joint Task
   Forces**.. 99

Bibliography.. 103

Figures

2.1.	JTF Headquarters by Service, 1990–Present	10
3.1.	Forming the Joint Planning Group: Example	25
3.2.	Army Combines Function-Based and Time-Based Elements	28
3.3.	Fifth Fleet Organization	31
3.4.	C-NAF Internal Structure	37
4.1.	Composition of JTF–Atlas Response HQ, 147 Total Personnel	45
4.2.	Composition of Combined Support Force–536 HQ, 986 Total U.S. Personnel	49
4.3.	Combined Support Force–536 Organizational Chart	51
4.4.	Composition of JTF–Noble Anvil HQ, 326 Total Personnel	56
4.5.	Composition of JTF–Southwest Asia HQ, 251 Total Personnel	61
7.1.	Comparison of a Notional AFFOR Staff with Other Typical Air Force Staffs	85

Tables

2.1.	Air Force–Led JTFs Since 1990	12
4.1.	Selected JTFs	41
A.1.	Joint Task Forces Since 1990	93
B.1.	Joint Manning Document for Joint Task Force Atlas Response Headquarters	99
B.2.	JTF/CSF—Unified Assistance (CSF-536) Headquarters	100
B.3.	Joint Task Force—Noble Anvil Headquarters	100
B.4.	Joint Task Force—Southwest Asia Headquarters	101

Summary

Since the late 1990s, the Air Force has made deliberate efforts to bolster its ability to effectively command and control air operations. These efforts have resulted in material and organizational changes to the air and space operations center (AOC) and an increase in the capability of the Joint Force Air Component Command (JFACC). At a higher echelon, recurring dissatisfaction with the approach to JTF command has led the Department of Defense (DoD) to call upon each of the services to be capable of fielding JTF headquarters.[2] To build effective JTF headquarters, commanders, and staffs, the Air Force will have to embark on a program similar to the one it did to build AOC and JFACC.

The Secretary of Defense or a combatant commander chooses JTF commanders. JTF headquarters plan and execute campaigns at the operational level of war. They take guidance from strategic-level authorities and combatant commanders and use it to shape missions. Then they allocate available means to undertake these missions. JTF headquarters, then, have two basic functions: planning and oversight of operations.

---

[2] U.S. Department of Defense, *Quadrennial Defense Review Report,* Washington, D.C., September 30, 2001, pp. 33–34; Donna Miles, "Core Elements Improve Crisis Response, Combat Ops," *American Forces Press Service,* March 23, 2006; U.S. Department of Defense, *Quadrennial Defense Review Report,* Washington, D.C., February 6, 2006a.

Of all the services, the Army is most frequently called upon to provide the core of JTF headquarters.[3] Air Force units have led at least 15 JTFs since 1990, but these have generally been rather small-scale noncombatant evacuations and humanitarian relief operations. The potential for air power to play larger roles in future conflicts suggests that the Air Force may need to be considered more often to lead future joint combat operations.

The Air Force should be prepared to supply JTF headquarters to the joint force when appropriate. It should identify and prepare units for this role. Operations that might best lend themselves to command by an airman might include those that are dominated by the use of land-based aircraft or those that take place across long distances. Likewise, Air Force personnel generally should not be considered for operations in which the predominance of forces are supplied by the Army, Marine Corps, or Navy. Of course, in many cases the choice of command will not be clear cut. By doing the best it can to generate competent commanders and staffs, the Air Force can be a more effective joint player, and it can better serve the nation.

This monograph surveys how the other services and other selected DoD organizations consider the issue of command and how their initiatives compare to similar efforts in the Air Force. We find that staffs balance between two different types of work: employment and management. We also find that staffs tend to be organized around one of two principal concerns: time and function. The Air Force will need to consider these factors as it considers its approach to the organization of future headquarters.

To help understand some of the issues involved in creating and operating JTF headquarters, this monograph examines four different JTFs. Two of them—JTF–Atlas Response (JTF-AR) and Combined Support Force (CSF)-536—were humanitarian operations, and two—JTF–Southwest Asia (JTF-SWA) and JTF–Noble Anvil (JTF-NA)—

---

[3] By *core of JTF headquarters,* we mean the commander and key elements of the headquarters staff. For the headquarters to reach full functionality, it needs to be augmented with additional staff from both the host service and the other services.

were combat operations. Two of them—JTF-AR and JTF-SWA—were led by Air Force units.

To create JTF headquarters, the Air Force must build them by selecting and molding commanders and staffs. It must also prepare to lead JTF headquarters by identifying the missions they may be assigned, by exercising commanders and staffs, and by engaging likely partners. In addition, the Air Force must execute or actually operate the JTF headquarters by building and maintaining partnerships, manning the headquarters, issuing orders, gaining and maintaining situational awareness, orchestrating efforts, and assessing and adjusting operations.

This analysis raises a number of questions for the Air Force including the following:

- Should Air Force JTF headquarters separate or combine employment and management tasks?
- Should Air Force JTF headquarters organize around time or function?
- When should Air Force units form the core of JTF headquarters and how many types of JTF headquarters does the Air Force need?
- How many JTF-capable numbered air force headquarters should the Air Force field?
- How would the Air Force simultaneously provide air component staffs and JTF headquarters?
- How should the Air Force staff JTF headquarters positions?
- How would the JTF headquarters incorporate other services and non-DoD partners?

Lastly, this monograph makes some suggestions for how the Air Force can increase its ability to form JTF headquarters. These recommendations fall under three categories: systems, people, and processes.

Systems

Acquire the necessary systems (pp. 65, 70) to send and receive information from fielded forces in the air, in space, at sea, and on land.

Determine the desired degree of "reach back" (p. 39). Assess which tasks may be best accomplished from a distance and how the forward command center will incorporate inputs from stations based in the continental United States. Determine how much *reach back* is necessary, possible, and desirable.

People

Reward those with deep experience in joint, interagency, and multinational operations (pp. 65–66). If the Air Force wants its C-NAFs to be capable of JTF leadership, it should provide incentives for officers to gain experience in working with partners outside the Air Force. By ensuring that officers who have spent more than one tour with other organizations are, in general, promoted at a rate equal to or above that of others, the Air Force can send a message that it seeks to develop well-rounded officers who have gained specific knowledge about military operations in other domains and about how other organizations work and more general lessons about how to establish effective working relationships with non–Air Force personnel.

Reorient professional military education (p. 85). Place more emphasis on planning in the curricula of key schools.

Assign competitive people to Air Force Forces (AFFOR) staffs (pp. 66–67). If the Air Force wants its C-NAFs to be capable of JTF leadership, and if it decides to staff JTF headquarters with AFFOR staffs, it needs to ensure that AFFOR staffs are populated by competent and respected personnel.

Train AFFOR staffs (pp. 66–67). Develop a training program to help staffs prepare for both AFFOR and JTF roles.

Processes

Designate JTF-capable organizations (pp. 75–76). In consultation with the designated C-NAFs and their respective combatant commands, the Air Force should specify general mission areas that C-NAF should be capable of undertaking.

Institute exercise programs (pp. 42–47, 68). Such a step would increase readiness for JTF headquarters duty and demonstrate this capability to combatant commanders.

Place more emphasis on planning (p. 70). Settle on an Air Force approach to operational planning that is applicable to both the air component and JTF headquarters roles, and teach this approach at Air Command and Staff College and at other appropriate venues.

Write a directive on Air Force JTF operations (p. 65). The directive would need to lay out how the Air Force as an institution and how individual AFFOR staffs would build JTF headquarters capability, and would task different Air Force entities to help make the vision of an Air Force JTF headquarters into a reality.

Learn JTF headquarters processes (pp. 69–72). Those who lead and man JTF headquarters need to know how to request forces from other services and how to issue formal orders to non–Air Force personnel.

Create the capability to deploy headquarters (pp. 24, 44). Other services have this capability, which the Air Force, in some cases, may need to replicate.

Create a champion for Air Force command (pp. 36–40). A recognized advocate for the key function of command would help to ensure that it is represented in debates over how to allocate resources.

Acknowledgments

The study team benefited tremendously from the advice and assistance of a number of people, some of whom cannot be named here. The idea for the study emerged from conversations with Maj Gen R. Michael Worden, then the HAF/A5X, and Andrew Hoehn, Vice President and Director of Project AIR FORCE. Both provided encouragement and support throughout the project. David Shlapak of RAND led the project in its early stages and was instrumental in structuring the study. Col Robert Evans and Gilbert Braun, both of HAF/A5XS, were generous with their time and views and helped shepherd the project through the Air Staff. Maj Gen (ret) John Corder and Benjamin S. Lambeth, both of whom have contributed significantly to the theory and practice of military operations, reviewed an earlier draft of this monograph with care and diligence.

Several retired Air Force officers—Lt Gen (ret) Joseph Hurd, Lt Gen (ret) Michael A. Nelson, and Lt Gen (ret) Michael C. Short—provided their views on the overall topic of JTF command as well as their experiences in JTFs. Lt Gen (ret) Joseph Hurd generously reviewed a draft outline of the monograph. Lt Col Michael "Starbaby" Pietrucha kindly reviewed an early draft while he was deployed to Iraq.

We would like to thank a number of other Air Force personnel, including Lt Col Todd Ackerman, Lt Col Vincent Alcazar, Lt Col Theodore Anderson, Col Brian Bartels, Col Chris Bence, Gen Roger A. Brady, Lt Gen Philip M. Breedlove, Lt Col Peter R. Brotherton, Col Mace Carpenter, Lt Col Eric Casler, Gen Carrol "Howie" Chandler, Stephen D. Chiabotti, Lt Col Jonathan Clough, Col Gary Crowder,

Lt Gen Daniel J. Darnell, Lt Gen David A. Deptula, Col Vincent DiFronzo, Maj Orlando Dona, Steven Dreyer, Lt Col Donald Finley, Brig Gen John C. Fobian, Julio C. Fonseca, James W. Forsyth, Lt Col Scot Gere, Dan Gnagey, Col Joe Guastella, Col Paul Harman, Wing Commander Tuben Harris (Royal Air Force), Lt Col Greg Hillebrand, Lt Col Clint Hinote, Lt Col Mark Hoehn, Maj Gen William Holland, Maj Gen James P. Hunt, Bill Jackson, Maj Gen Kevin Kennedy, Col Michael Korcheck, Lt Col Lori LaVezzi, Maj Doug Lee, Brooks Lieske, Gen Stephen Lorenz, Renee Maisch, Jim McDonnell, Brian Mclean, Sydney McPherson, Bill Minich, Col John Murphy, Col (ret) Matt Neuenswander, Lt Gen Allen Peck, Brig Gen Robin Rand, Maj Gen William Rew, Phillip M. Romanowicz, Lt Gen (ret) Eugene Santarelli, Lt Col Richard A. Seifert, Robert Sligh, Michael Smellie, Lt Col Jeff Smith, Col Murrell Stinnette, Lt Gen Loyd Utterback, Maj Gen Richard E. Webber, and Maj Tony Zanca. These individuals offered their views and shared information and insights about how the Air Force approaches the issue of command. Maj Michael Nelson went out of his way to provide assistance on an extremely useful trip to the Ninth Air Force facility in the Middle East.

A number of Army personnel offered assistance, including LTC Bryan G. Cox, LTC Telford Crisco, Lou Gelling, MAJ Michael Halley, COL Glenn W. Harp, MAJ Dennis Malone, LTC James McFadden, COL Eric Nelson, LTC Thomas R. Taylor, CW4 Jacqueline Wallace, and COL Stuart Whitehead.

From the Navy, we would like to thank CAPT M. J. Barea, CAPT Joe Bauknecht, LCDR Anthony Butera, CDR Shan Byrne, CDR Lynn Chow, VADM Kevin J. Cosgriff, CDR Keith Holden, CAPT Hank Miranda, CDR Joe Polanin, CAPT Mike Smacker, CAPT Mike Spence, and CAPT Steve Zaricor. LT Ananda Mason provided critical assistance for an informative trip to U.S. Naval Forces Central Command.

We were fortunate to have the help of John Bacon, Col. David Garza, Antonio Mattaliano, and Marty Westphal of the Marine Corps. At Joint Force Command, we would like to thank Dewey Blyth, Bill Brown, Col Donnie Davis, Col Paul Haveles, Bill Newlon, and Mike Rapp.

At RAND, we would like to thank Irv Blickstein, Tim Bonds, Cynthia Cook, Natalie Crawford, John Crown, Paul Emslie, Leland Joe, David Johnson, Wade Markel, Kimbria McCarty, Ron Miller, Karl Mueller, Walt Perry, Christina Pitcher, Al Robbert, Brian Shannon, and Peter Soriano. David Ochmanek provided wisdom and thoughtful consideration of several drafts and ideas. Skip Williams offered kind advice, and Molly Coleman supplied administrative support and perspective.

Abbreviations

ACOS	Assistant Chief of Staff
ADCON	administrative control
AFFOR	Air Force Forces
AFIC	Armed Forces Inauguration Committee
AFRICOM	Africa Command
AOC	air and space operations center
AOR	area of responsibility
ARFOR	Army forces
ATO	air tasking order
AVN	aviation
BCTP	Battle Command Training Program
BPZ	below primary zone
C2	command and control
C4	command, control, communications, and computers
C5F	Fifth Fleet
CA	Civil Affairs

CADRE	College of Aerospace Doctrine, Research, and Education
CDR	commander
CDRUSAPACOM	Commander, United States Pacific Command
CE	Core Element
CENTCOM	(United States) Central Command
CFACC	Combined Force Air Component Command
CFLCC	Combined Force Land Component Command
CFMCC	Combined Force Maritime Component Command
CIV	civilian
CJCS	Chairman of the Joint Chiefs of Staff
CJTF	combined joint task force
CMD	command
CMOC	civil-military operations center
C-NAF	component–Numbered Air Force
COMAFFOR	commander of Air Force Forces
CONOPLAN	Concept of Operations Plan
CONOPS	concept of operations
COPS	current operations
COS	Chief of Staff
CP	command post
CSF	Combined Support Force
CSG	Combined Support Group

CSG-IN	Combined Support Group–Indonesia
CSG-SL	Combined Support Group–Sri Lanka
CSG-TH	Combined Support Group–Thailand
CSSC	COMAFFOR Senior Staff Course
CTF	combined task force
CUSNC	Commander, U.S. Naval Forces Central Command
DART	Disaster Assistance Response Team
DCFMCC	Deputy Combined Force Maritime Component Commander
DCOM	deputy commander
DCUSNC	Deputy Commander, U.S. Naval Forces Central Command
DIRSPACEFOR	Director of Space Forces
DJFMCC	Deputy Joint Force Maritime Component Commander
DJIOC	Defense Joint Intelligence Operations Center
DNA	Defense Nuclear Agency
DoD	Department of Defense
DRAT	Disaster Response Assessment Team
ENGR	engineer
EPOC	(United States) European Command Plans and Operations Center
EUCOM	(United States) European Command
FDO	flexible deterrent options

FltMgmt	fleet management
FOPS	future operations
FPC	future plans
FRAGO	Fragmentary Order
FRY	Former Republic of Yugoslavia
FSC	fire support coordination
G5	Plans
HA	humanitarian assistance
HN	humanitarian
HOA	Horn of Africa
HQ	headquarters
HUMRO	humanitarian relief operation
IAG	Iraq Assistance Group
IDE	intermediate developmental education
IO	international organization
ISAF	International Security and Assistance Force
ISR	intelligence, surveillance, and reconnaissance
ITFC	Iraq Threat Finance Cell
JAC	Joint Analysis Center
JAOP	joint air and space operations plan
JCS	Joint Chiefs of Staff
JFACC	Joint Force Air Component Command
JFCOM	(United States) Joint Force Command
JFLCC	Joint Force Land Component Command

JFMCC	Joint Force Maritime Component Command
JHQ	joint headquarters
JIB	Joint Information Board
JMD	joint manning document
JMEP	Joint Manpower Exchange Program
JOPES	Joint Operation Planning and Execution System
JP	joint publication
JPG	joint planning group
JSOTF	joint special operations task force
JTF	joint task force
JTF-AR	Joint Task Force–Atlas Response
JTF-NA	Joint Task Force–Noble Anvil
JTF-SWA	Joint Task Force–Southwest Asia
JTTR	Joint Theater Trauma Registry
LCC	logistics coordination cell
LNO	liaison officer
MAJCOM	major command
MCO	major combat operations
MCP	main command post
MDMP	Military Decision Making Process
MEF	Marine expeditionary force
MHQ	Maritime Headquarters
MNC-I	Multi-National Corps–Iraq
MNF-I	Multi-National Force–Iraq

MOC	Maritime Operations Center
MOCDIR	Maritime Operations Center director
MOS	military occupational specialty
MPAT	Multinational Planning Augmentation Team
MSTP	Marine Air-Ground Task Force Staff Training Program
NAF	Numbered Air Force
NATO	North Atlantic Treaty Organization
NAVCENT	(United States) Naval Forces Central Command
NAVFOR	Navy forces
NEO	noncombatant evacuation operation
NFZ	no-fly zone
NGO	nongovernmental organization
NSJ	National Scout Jamboree
NSSE	National Special Security Event
NTIS	National Technical Information Service
OAF	Operation Allied Force
OCP	operational command post
OFDA	Office of Foreign Disaster Assistance
OMA-A	Office of Military Assistance
ONA	operational net assessment
OPCON	operational control
OPs	operations
OpsSpt	operations support

OSCA	Office of Security Cooperation Afghanistan
OSF	operations support facility
PACAF	Pacific Air Forces
PACOM	United States Pacific Command
PAD	Program Action Directive
PAF	Project AIR FORCE
PAO	Public Affairs Office
PSO	peace support operation
PSYOPS	psychological operations
PVO	private volunteer organization
QDR	Quadrennial Defense Review
REQ	required
RFF	Request for Forces
SAASS	School of Advanced Air and Space Studies
SAMS	School of Advanced Military Science
SASO	stability and support operation
SDE	senior developmental education
SGM	sergeant major
SJFHQ	Standing Joint Force Headquarters
SOCCENT	Special Operations Command Central
SOCPAC	Special Operations Command Pacific
SOF	special operations forces
TAC CP	tactical command post
TACON	tactical control

TLAM	Tomahawk Land-Attack Missile
UN	United Nations
UNSC	United Nations Security Council
UNSCR	United Nations Security Council Resolution
USAF	United States Air Force
USAFE	United States Air Forces, Europe
USAID	United States Agency for International Development
USEUCOM	United States European Command
USGA	United States government agency
USJFCOM	United States Joint Force Command
USMC	United States Marine Corps
USN	United States Navy
USNAVCENT	United States Naval Forces Central Command
USPACOM	United States Pacific Command
UTC	unit type code
VCUSNC	Vice Commander, U.S. Naval Forces Central Command

CHAPTER ONE

Introduction and Purpose

The U.S. Air Force has devoted considerable resources toward building its component headquarters, with good reason: These organizations develop air campaign plans and manage their execution during crises and conflicts. But many operations are led by joint task force (JTF) commanders, and prominent members of the defense community have called upon each of the services to be prepared to provide JTF headquarters.[1] The purpose of this report is to help the Air Force better understand the role and demands of JTF headquarters. To play a leading role in such operations, the Air Force will need a good understanding of what it takes to provide overall command of them. Air Force leaders and policy papers have stated that Air Force components of combatant commands will be able to serve as JTF headquarters if called upon to do so. However, beyond these intentions there has been relatively little consideration within the service about how the Air Force should go about providing JTF commanders and staffs. Granted, there have been some instances in which Air Force units have provided the core of JTF headquarters. However, these have tended to be the exception rather than the rule, and the Air Force could do considerably more to prepare its people for this role.

Accordingly, this monograph also seeks to identify the broad outlines of a way forward for the Air Force to best prepare commanders

---

[1] U.S. Department of Defense, *Quadrennial Defense Review Report,* Washington, D.C., September 30, 2001, pp. 33–34; Donna Miles, "Core Elements Improve Crisis Response, Combat Ops," *American Forces Press Service,* March 23, 2006; U.S. Department of Defense, *Quadrennial Defense Review Report,* Washington, D.C., February 6, 2006a.

and staffs to lead JTF headquarters. The Air Force should be prepared to supply JTF headquarters to the joint force when appropriate. Operations dominated by the use of land-based aircraft or which take place across long distances are obvious candidates for Air Force leadership. On the other hand, the Air Force should not be considered for operations in which the predominance of forces are supplied by the Army, Marine Corps, or Navy. However, there are murkier examples, such as operations that include significant land or naval elements as well as land-based aviation units. By doing the best it can to develop competent commanders and staffs, the Air Force can become a more effective participant in joint operations and, in so doing, better serve the nation.

Chapter Two of this monograph defines JTFs in theory and practice. It discusses why providing JTF headquarters is an issue for the Department of Defense (DoD), and why it is an objective for the Air Force. Chapter Three considers some general themes in current U.S. military conceptions of command, and it surveys some command initiatives being undertaken or considered by other U.S. military services and DoD entities. Chapter Four examines four JTF headquarters from recent contingencies to derive implications for future commands. Chapter Five considers the requirements for JTF headquarters, and Chapter Six reviews issues that the Air Force will wish to consider as it prepares to provide this capability. The monograph closes with a set of recommendations for the Air Force to provide JTF headquarters, focusing on three areas: systems, people, and processes. Two appendixes provide data on past JTFs and joint manning documents from selected JTFs.

CHAPTER TWO
Background

JTFs in Theory

Joint doctrine is rather vague about the purpose of the JTF. It states that a "joint task force is a joint force that is constituted and so designated by a JTF establishing authority."[1] In other words, a JTF is whatever the Secretary of Defense, combatant commander, subordinate unified commander, or existing JTF commander says it is. This definition does point out that the JTF is "joint"—it, therefore, has authority over forces from more than one service.

Commanding authorities tend to choose JTF commanders and staffs from the service components associated with the area of responsibility (AOR) of a particular regional combatant command. Joint doctrine suggests that this is the preferred option for establishing a JTF headquarters.[2] There is considerably less guidance in joint doctrine relating to the question of what criteria JTF establishing authorities should use when assigning a unit to be a JTF headquarters. One basic assumption is that the nature of the operation and mission requirements should inform choices about which service should provide the JTF headquarters, and that the service with the "preponderance of forces" will likely be asked to lead the operation.[3] For example, if the mission requires mostly ground forces, it is likely that an Army or

---

[1] Joint Publication 3-33, *Joint Task Force Headquarters,* Washington, D.C.: The Joint Staff, February 16, 2007, p. I-1. See also Joint Publication 3-0, *Joint Operations,* Washington, D.C.: The Joint Staff, September 17, 2006, Chapter II.

[2] Joint Publication (JP) 3-33, 2007, p. II-1.

[3] See the discussion in JP 3-33, 2007, p. II-2.

Marine Corps unit will be assigned as the JTF headquarters.[4] Using "preponderance" as a criterion is not without its problems, however. For instance, it is not always clear which service has the "most" forces committed to an operation. One tank does not equal one plane or one ship. Numbers of personnel can also be deceiving. One might argue that the service that produces the forces that are most vital to the operation's success should command the JTF, but this may also require subjective judgment.

Others note that the unit selected should have the capability to command and control forces involved in the operation. This should certainly be a necessary condition for selecting a unit. If a headquarters cannot communicate with fielded forces, it cannot purport to lead them. Nevertheless, the ability to command and control is not by itself a sufficient reason to choose one unit over another.

Another issue deserves mention before going further. Much writing in the defense community uses the terms "JTF" and "JTF headquarters" interchangeably. Strictly speaking, "JTF" refers to the entire force, encompassing the headquarters *and* the line and support units subordinate to it. Our concern in this study is primarily the headquarters element, which is composed of the commander and his or her staff. We will refer to the "JTF headquarters" when discussing the headquarters, and we will only use the term "JTF" when referring to the entire force. The commander leads both the JTF and the JTF headquarters. The commander exercises command and control of fielded forces through the headquarters.

A JTF headquarters plans and executes campaigns at the operational level of war.[5] JTF commanders receive guidance from superiors,

---

[4] This expectation does not always hold true. For example, in Operation Allied Force (OAF), which we will discuss in detail later in this monograph, the combatant commander created a JTF led by a Navy admiral and a Navy staff even though the operation was carried out primarily by land-based aircraft.

[5] Joint Publication 1-02, *The Department of Defense Dictionary of Military and Associated Terms*, Washington, D.C.: The Joint Staff, 2007, p. 394, defines the *operational level of war* as

> The level of war at which campaigns and major operations are planned, conducted, and sustained to achieve strategic objectives within theaters or other operational areas. Activ-

such as the president and the secretary of defense, about U.S. national interests at stake in the contingency and about the strategic ends they seek. These commanders distill this guidance into relatively discrete operational goals that can be reached using the means available. They must also craft a campaign plan and coordinate it, along with a set of associated rules of engagement, with their superiors and the military services providing forces. In addition, JTF commanders allocate these forces assigned to the operation as appropriate to meet these goals. This process involves allocating forces and setting priorities for their use. As the DoD definition of a JTF headquarters notes, JTF commanders have at their disposal capabilities provided by more than one service. Commanders, through their staffs, integrate and orchestrate these capabilities in the manner that they judge will give JTFs the best chances of meeting their goals.

JTF headquarters have two basic functions: planning and oversight of operations. JTF commanders and staffs plan either for specific missions or for a limited range of likely missions. Under the guidance given to them from national command authorities, they survey the operational environment and craft detailed courses of action to further U.S. interests. They integrate forces from two or more services into a common effort and are often called upon to incorporate allied or coalition forces into the operation. Effective JTF commanders and staffs prepare their forces to deal with a variety of changes in the operation. They consider "branches and sequels," or how they might respond to changing circumstances. These functions are often referred to collectively as "command."[6]

---

ities at this level link tactics and strategy by establishing operational objectives needed to achieve the strategic objectives, sequencing events to achieve the operational objectives, initiating actions, and applying resources to bring about and sustain these events.

JP 1-02, 2007, p. 76, defines a *campaign* as "a series of related major operations aimed at achieving strategic and operational objectives within a given time and space."

[6] JP 1-02, 2007, p. 101, defines *command* as

1. The authority that a commander in the Armed Forces lawfully exercises over subordinates by virtue of rank or assignment. Command includes the authority and responsibility for effectively using available resources and for planning the employment of, organizing, directing, coordinating, and controlling military forces for the accomplish-

JTF commanders and staffs also orchestrate and oversee operations. They take in information from sensors and other fielded forces to ensure that fielded forces are prosecuting the operation as closely as possible to its design. This function, which has more to do with implementation than with planning, is sometimes referred to as "control."[7] Some might prefer the term "execution."[8]

JTFs in Practice

Under normal conditions, the staffs at geographic combatant commands focus primarily on steady-state efforts: forging relationships

---

ment of assigned missions. It also includes responsibility for health, welfare, morale, and discipline of assigned personnel.

2. An order given by a commander; that is, the will of the commander expressed for the purpose of bringing about a particular action.

3. A unit or units, an organization, or an area under the command of one individual.

For other assessments of the nature of command, see Martin van Creveld, *Command in War*, Cambridge, Mass.: Harvard University Press, 1985; Carl H. Builder, Steven C. Bankes, and Richard Nordin, *Command Concepts: A Theory Derived from the Practice of Command and Control*, Santa Monica, Calif.: RAND Corporation, MR-775-OSD, 1999; and Kenneth Allard, *Command, Control and the Common Defense*, Washington, D.C.: National Defense University Press, revised edition 1996.

[7] This conception runs counter to the definition of *control* in JP 1-02, 2007, but we would argue that it is more useful. JP 1-02 (2007, p. 120) defines *control* as

1. Authority that may be less than full command exercised by a commander over part of the activities of subordinate or other organizations.

2. In mapping, charting, and photogrammetry, a collective term for a system of marks or objects on the Earth or on a map or a photograph, whose positions or elevations (or both) have been or will be determined.

3. Physical or psychological pressures exerted with the intent to assure that an agent or group will respond as directed.

4. An indicator governing the distribution and use of documents, information, or material. Such indicators are the subject of intelligence community agreement and are specifically defined in appropriate regulations.

[8] We are indebted to Maj Gen (ret) John Corder for this insight.

with partner militaries, revising war plans, and collecting information about their AOR. Often, they do not have the capacity to devote staff resources to command and control forces in contingencies, especially when multiple contingencies are taking place simultaneously in their AOR. Frequently, they have constituted JTF headquarters as a way to provide leadership for such operations. There is a lower and upper bound for contingencies that call for the creation of JTF headquarters. Below a certain level, a combatant command staff can handle an operation on its own without creating a JTF structure. At the other end of the spectrum, combatant commanders have tended to handle very large-scale operations on their own, proving reluctant to pass this responsibility to a subordinate. In this case, the combatant command staff tends to put aside day-to-day efforts to devote themselves to the crisis at hand.

Between these two poles lies a wide range of operations in terms of intensity, scope, and mission type. JTF headquarters have been established to provide leadership for a number of different missions, including counternarcotics efforts, noncombatant evacuation operations (NEOs), counterinsurgency, counterterrorism, humanitarian assistance (HA), military support to civilian authorities, security for special events, and combat operations.

The first JTF we have been able to identify was Joint Task Force One (JTF-1). On the same day in October 1945 that the Joint Chiefs of Staff (JCS) referred the issue of creating a Department of Defense to the President of the United States, the JCS received a recommendation that surplus U.S. ships and surrendered Japanese and German ships be used in nuclear weapons tests in order to determine the power of the atom bomb.[9] JTF-1 was organized on January 11, 1946, as part of Operation Crossroads at Bikini Atoll. Its mission was to increase understanding of the effects of nuclear weapons on ships and other equipment. The USS *Mount McKinley* housed the JTF headquarters. VADM W.H.P. Blandy, USN, led the JTF. Statements documenting

---

[9] James F. Schnabel, *History of the Joint Chiefs of Staff*, Vol. 1, *The Joint Chiefs of Staff and National Policy, 1945–1947*, Washington, D.C.: Office of Joint History, Office of the Chairman of the Joint Chiefs of Staff, 1996, p. 131.

the creation of JTF-1 are few, but in the presence of bitter fighting between the Army and Navy, particularly over responsibilities in the Pacific theater, the idea of a JTF, formed for the execution of a specific mission without reference to the military's larger command structure, might have seemed an appealing circumvention. When it was created, JTF-1 was specifically required to report directly to the JCS with a separate board to evaluate the tests' results. The decision to create the JTF, however, had to be approved by President Truman.[10] The necessity of presidential approval and oversight indicates the singular nature of this enterprise, rather than the expectation that the JTF concept was to become routine. However, military observers did recognize that Operation Crossroads would have ramifications for interservice rivalry; it was generally believed that the outcome of the tests would either prove the Navy's fleet resilient or crown the Army Air Force America's first line of defense.[11] The first bomb was dropped slightly off target from a B-29. It destroyed five ships and produced minimal amounts of radioactivity. The second detonated underwater, destroyed eight ships, and contaminated much of the target fleet.

Approximately 3,300 Army, 2,500 Army Air Force, 580 Marine Corps, and 37,000 Navy personnel composed JTF-1. Included in this group were some 501 Navy officers, 8 Marine Corps officers, 141 Army ground officers, and 21 Army Air Force officers. The vast majority of Navy officers, 444 in all, were assigned to the ships at Bikini Atoll, meaning that the commander's staff was less overwhelmingly naval in composition.[12]

While unification of command was a process that would take many years, the JTF has been used as an operational workaround of sorts. But because of the ad hoc nature of the JTF, it developed some-

---

[10] JCS 1552/6, December 29, 1945; Memo, JCS for SecWar and SecNavy, "Tests of the Effects of Atomic Explosives," December 1945, cited in Schnabel, 1996, p. 227.

[11] Lloyd J. Graybar, "The 1946 Atomic Bomb Tests: Atomic Diplomacy or Bureaucratic Infighting?" *The Journal of American History*, Vol. 72, No. 4, March 1986, p. 894.

[12] L. Berkhouse, F. W. McMullan, S. E. Davis, and C. B. Jones, *Operation Crossroads—1946: United States Atmospheric Nuclear Weapons Tests*, Nuclear Test Personnel Review, Washington, D.C.: Defense Nuclear Agency, NTIS, DNA 6032F, 1984, pp. 160, 164, 172, 188, 206.

what organically, without a great deal of forethought or structure. This type of development has allowed for a great deal of flexibility, but little consistency or standardization.

As we can see from JTF-1, the JTF headquarters represents an early attempt to minimize the impact of interservice rivalry on operations. It has served this purpose fairly well over the years. While a service-based unit is often designated as a JTF headquarters, it is expected to bring in personnel from other services to help provide expertise across the range of mediums, platforms, and capabilities needed for the conduct of the operation. A service headquarters can augment itself in three ways. It can request personnel through the Joint Manpower Exchange Program (JMEP), which assigns officers from one service to work in another service's component headquarters.[13] It can also turn to the Standing Joint Force Headquarters (Core Element) or SJFHQ (CE), a standing joint team based at each combatant command that is trained and equipped to advise JTF headquarters and to help staff them.[14] In addition, a service component headquarters that takes the role of a JTF headquarters can request additional personnel through a joint manning document (JMD).[15] The unit must have the combatant commander and Joint Staff approve the request. Then, the combatant commander formally requests the forces from the services. Next, the services appoint individual augmentees to fill positions. As we will see below, despite the JMD process, JMEP, and SJFHQ (CE), JTF headquarters often find it difficult to fill required staff slots.

---

[13] Jennifer Colaizzi, "Positive Review for Joint Manpower Exchange Program," *USJFCOM Public Affairs*, September 21, 2005.

[14] COL Douglas K. Zimmerman, USA, "Understanding the Standing Joint Force Headquarters," *Military Review*, July–August 2004; and U.S. Joint Force Command, "Standing Joint Force Headquarters (Core Element)," briefing, COL Douglas K. Zimmerman, USA, January 5, 2007. There are also two SJFHQs (CE) based at the U.S. Joint Force Command (JFCOM).

[15] Guidance for the development of JMDs and augmentation of unfunded personnel to a JTF headquarters is provided in Chairman of the Joint Chiefs of Staff, "Individual Augmentation Procedures," Instruction, Washington, D.C.: The Joint Staff, CJCSI 1301.01C, January 1, 2004.

Combatant commanders have frequently turned to JTF headquarters to handle operations. There have been over 300 JTFs since 1970.[16] Appendix A lists JTFs that have been in operation since 1990.[17] Figure 2.1 shows that of the 90 JTF headquarters constituted between 1990 and 2008, Army units have formed the core of 47 (52 percent), while the Air Force has formed the core of 15 (17 percent). The Navy has formed the core of 11 (12 percent), and 9 have been led by the U.S. Marine Corps (USMC) (10 percent).[18] The remaining 8 were either led by subcomponents of combatant commands, leadership was rotated among the services, or the leadership is unknown.

Figure 2.1
JTF Headquarters by Service, 1990–Present

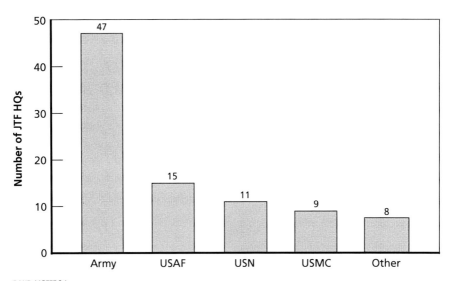

RAND MG777-2.1

---

[16] Timothy Bonds, Myron Hura, and Thomas-Durell Young, *Enhancing Army Joint Force Headquarters Capabilities,* Santa Monica, Calif.: RAND Corporation, MG-675-A, forthcoming.

[17] Unfortunately, there is no single official source of data on JTFs and their headquarters.

[18] JTF-160, JTF-510, and Joint Contracting Command Iraq/Afghanistan rotated their leads among services.

Upon closer inspection, Air Force–led JTFs have not been assigned the most demanding missions. Table 2.1 lists Air Force–led JTFs since 1990. Of the 15 Air Force–led JTF headquarters, 4 were humanitarian relief operations (HUMROs), 2 were to help evacuate noncombatant evacuees, and 4 provided support to other forces. The 5 "combat" JTFs enforced no-fly zones (NFZs) in northern and southern Iraq and Bosnia and conducted an air campaign from Turkey during Operation Desert Storm. While the Air Force has led JTFs, it has never led one of considerable scale or complexity or one that has required it to synchronize the efforts of sizable elements of forces from different mediums.

The Problem

DoD leaders have expressed concern that the ad hoc nature of JTF headquarters has reduced their effectiveness.[19] In building its JTF headquarters capability, the Air Force will have to address this concern. DoD has experienced some difficulties in fielding JTF headquarters, and, as a consequence, senior defense officials have expressed a need for more capability in this area. Newly constituted JTFs confront a number of challenges. Often, they are given little or no warning before being directed into action. When a crisis erupts that calls for the creation of an operational-level commander and staff, there is little time to plan or ponder courses of action. For example, Joint Task Force–Atlas Response (JTF-AR), one of the JTFs that we examine later in this monograph, had no time between the first order relating to the operation and the execution order for the operation. One study found that 68 percent of all JTF headquarters had less than six weeks to prepare for operations.[20]

While they start quickly, JTFs and their associated headquarters generally persist for longer periods of time than has been the case in the past. During the 1970s, JTFs lasted an average of 73 days. This

---

[19] Miles, 2006.

[20] Armando X. Estrada, *Joint Task Force Requirements Determination: A Review of the Organization and Structure of Joint Task Forces*, Monterey, Calif.: Naval Postgraduate School, Graduate School of Business and Public Policy, March 2005, p. 31.

Table 2.1
Air Force–Led JTFs Since 1990

Name	Mission	Start	End	Type of Mission	Mission Details
JTF–Proven Force	Combat air operations (Iraq)	1990	1991	Combat	USAFE-led air combat operations from Incirlik against Iraq in Operation Desert Storm
JTF–Fiery Vigil	Humanitarian evacuation (Philippines)	1991	1991	NEO	Thirteenth Air Force–led evacuation of nonessential personnel and dependents from Clark Air Force Base in the Philippines following the eruption of Mt. Pinatubo
JTF–Quick Lift	NEO (Zaire)	1991	1991	NEO	USAFE-led JTF that deployed French and Belgian troops to Zaire and evacuated 716 people following a mutiny
JTF–Provide Hope	HA (Soviet Union)	1992	1992	HUMRO	USAF Military Airlift Command–led humanitarian airlift operation to the Soviet Union
JTF–Provide Transition	Support (Angola)	1992	1992	Support	USAFE-led airlift operation (with 326 sorties) to relocate government and rebel soldiers in Angola in support of democratic elections
JTF–Southwest Asia	NFZ enforcement (Iraq)	1992	2003	Combat	Execution of an NFZ below the 32nd parallel in Southern Iraq led by U.S. Central Command Air Forces
JTF–Deny Flight/Decisive Edge	NFZ enforcement (Bosnia/Herzegovina)	1993	1995	Combat	NATO enforcement of the NFZ over Bosnia, providing close air support to UN troops, and conducting approved air strikes under a "dual-key" command arrangement with the UN
JTF–Deliberate Force	Air campaign (Bosnia)	1995	1995	Combat	Oversee the air campaign against Serbian forces in Bosnia/Herzegovina

Table 2.1—Continued

Name	Mission	Start	End	Type of Mission	Mission Details
JTF–Pacific Haven	Refugee screening (Guam)	1996	1997	HUMRO	Humanitarian airlift and support of Kurdish refugees from Iraq to Incirlik and then to Guam
CTF–Northern Watch	NFZ enforcement	1997	2003	Combat	Execution of an NFZ above the 36th parallel in Northern Iraq led by USAFE
JTF–Assured Lift	NEO/Movements (Liberia)	1997	1997	Support	Provide airlift and logistical support to West African peacekeepers deployed to Liberia
JTF–Eagle Vista	Other	1998	1998	Support	Support for the U.S. president's visit to Africa
JTF–Shining Hope	Humanitarian relief (Kosovo)	1999	1999	HUMRO	Provide humanitarian relief to Kosovars in Albania and Macedonia
JTF–Atlas Response	Humanitarian relief (Mozambique)	2000	2000	HUMRO	Humanitarian airlift in support of international relief effort in response to massive floods in Mozambique
JTF–Autumn Return	NEO (Cote d'Ivoire)	2002	2002	NEO/SOF	NEO from Cote d'Ivoire, which included forces from the 352nd Special Operations Group

increased to an average of 117 days in the 1980s, and then to 374 days in the 1990s. As of 2005, JTF duration averaged 637 days.[21]

Growth in the period of JTF operations reflects the high operational tempo that U.S. forces have had to bear since the end of the Cold War. Longer-lasting JTFs place greater strain on equipment and manpower. The need to man JTF headquarters has placed strain on a wide range of personnel. People selected to staff JTF headquarters are

---

[21] Bonds, Hura, and Young (forthcoming); Estrada (2005); and U.S. Joint Force Command, *Expanding the Joint C2 Capability of Service Operational Headquarters*, Strategic Planning Guidance 2006–2011 Directed Study Task 04, February 17, 2005.

taken away from their normally scheduled duties, which means that for the duration of each JTF, other tasks across the U.S. armed forces are either left undone or undertaken by fewer people than would otherwise be the case. In addition, positions through the JMD process are unfunded, which means that the services continue to pay the costs of their personnel while they conduct work for the JTF.

For these reasons, JTF headquarters often find it difficult to reach their authorized manning levels. The services are mindful of the value of the people that they might send to JTF headquarters and are often reluctant to let them go. For example, at the end of Operation Allied Force (OAF), the JTF headquarters still had 20 percent of its positions empty.[22] The number of officers assigned through JMEP is rather small compared with the needs of a JTF headquarters. SJFHQ (CE) is also rather small, consisting of fewer than 60 personnel. The JMD process is notoriously slow. It can take up to six months to gain JCS approval for a JMD, and even after approval, the services have been reluctant to send personnel to work in JTF headquarters.[23] These problems contribute to the perception that, despite their designation as "joint" entities, JTF headquarters tend to be dominated by one service.

The wide variety of missions assigned to JTF headquarters presents another challenge. As noted above, JTF headquarters have been created to oversee combat, peacekeeping, stability and security, humanitarian support, counterdrug, and noncombatant evacuation operations. They have also conducted training and have provided security for special events such as the Olympics and the Boy Scout Jamboree. With such a variety of mission types, it is difficult for would-be JTF commanders and staffs to organize, train, and plan effectively in advance.

In addition to these difficulties, JTF headquarters are generally created to handle vexing issues that affect U.S. national interests. If the problem were easily solved, it would not be necessary to set up a JTF headquarters. If the tasks were not considered to be important, the

---

[22] Jim Garamone, "QDR Approves Joint Force Headquarters Concept," *American Forces Press Service*, October 29, 2001.

[23] Bonds, Hura, and Young (forthcoming).

United States would not expend resources sourcing the JTFs and their headquarters.

DoD has attempted to redress these difficulties. In the first Quadrennial Defense Review (QDR) issued during his tenure, Secretary of Defense Donald Rumsfeld called for the establishment of an SJFHQ at each regional combatant command.[24] Later, Secretary Rumsfeld declared that the nascent SJFHQs would provide "more standing joint force capability so that we don't have to start from a dead start."[25] An office was established at the U.S. Joint Force Command (JFCOM) to build and promote the use of SJFHQs. The vision for SJFHQ has yet to be fulfilled, because proponents have faced difficulties in finding required levels of manpower and in producing more-uniform organizations in different combatant commands. The 2006 QDR highlighted the need to provide more-effective command capabilities for contingencies, but it called upon the services to play a greater role:

> Rapidly deployable, standing joint task force headquarters will be available to the Combatant Commanders in greater numbers to meet the range of potential contingencies. . . . [DoD will] transform designated existing Service operational headquarters to fully functional and scalable Joint Command and Control Joint Task Force–capable headquarters beginning in Fiscal Year 2007.[26]

This vision is similar to the drive within the Air Force to reduce the ad hoc nature of the air and space operations center (AOC) and to produce more standardized Air Force components for combatant commands. As we will see, recent Air Force efforts have focused on the component level. To fulfill the direction of the 2006 QDR, the Air Force will have work to do.

---

[24] U.S. Department of Defense, 2001, pp. 33–34.
[25] Miles, 2006.
[26] U.S. Department of Defense, 2006a.

The Objective

Many in the Air Force would like to see more Air Force officers serve as JTF commanders and to have Air Force organizations serve as the core of JTF headquarters staffs. They argue that Air Force personnel have a unique, theaterwide perspective due to the relative speed with which air platforms traverse the battlespace, and that this perspective can help increase the performance of the joint force.[27] Two RAND analysts argue that the Air Force's performance in recent conflicts indicates that the service has increased its ability to contribute to the joint fight.[28] This increase in capability, coupled with the nature of U.S. national security challenges, has led some to call for a shift in the roles the services play in conflict, with the Air Force and Navy taking more responsibility for large-scale conflict and portions of the Army and Marine Corps focusing more on stability operations.[29]

We showed above that the Army provides more JTF headquarters than other services. It is often the nation's "supported" service and the leader of the joint force.[30] The Army places emphasis on preparing its officers to craft detailed plans for force employment. When asked to consider the possibility of Air Force–led JTFs, Army officers are often skeptical. They argue that the Air Force does not understand ground combat and the art of maneuver and could not credibly lead a signifi-

---

[27] Rebecca Grant, looking at the question of why more Air Force personnel have not been selected to lead regional combatant commands, argues that "if air power is the dominant force in today's military operations—and it is—you would expect to see more airmen in command." Rebecca Grant, "Why Airmen Don't Command," *Air Force Magazine,* March 2008, pp. 46–49.

[28] David E. Johnson, *Learning Large Lessons: The Evolving Roles of Ground Power and Air Power in the Post–Cold War Era*, Santa Monica, Calif.: RAND Corporation, MG-405-1-AF, 2007; Benjamin S. Lambeth, *The Transformation of American Air Power*, Ithaca, N.Y.: Cornell University Press, 2000.

[29] Andrew R. Hoehn, Adam Grissom, David A. Ochmanek, David A. Shlapak, and Alan J. Vick, *A New Division of Labor: Meeting America's Security Challenges Beyond Iraq*, Santa Monica, Calif.: RAND Corporation, MG-499-AF, 2007.

[30] For more discussion on this point, see John Gordon IV and Jerry Sollinger, "The Army's Dilemma," *Parameters,* Summer 2004.

cant joint fight.³¹ Some also argue that the careers of most Air Force officers are too narrowly focused on air power and that officers do not tend to have much exposure to forces operating in other mediums or to other services or organizations.

Air Force officers have rarely led joint forces in theater-level combat or other major operations.³² This is partly the unavoidable result of history and technological progress: Armies and navies have existed for millennia. Powered flight, on the other hand, is just over 100 years old, and the Air Force is celebrating its 60th year of being an independent service. Hence, some skepticism about Air Force officers' ability to command large-scale, joint operations is to be expected. The burden of proof rests on the Air Force to demonstrate its ability and intention to play more of a leadership role in joint operations.

Skepticism about the desirability of Air Force leadership of JTF headquarters also raises the question of what sorts of contingencies would be most appropriate for Air Force–led JTF headquarters. Air Force officers have successfully led HUMROs in the past, and the lift, sensor, and communications capabilities that the service provides suggest that this is a natural mission for Air Force leadership. The same capabilities make NEOs another obvious mission.

It would be less appropriate for an Air Force organization to provide the commander and staff for operations that require large numbers of ground troops and less emphasis on air and space platforms, such as most peacekeeping, counterinsurgency, or stability and support operations. This is not to say that the Air Force does not make significant contributions to these types of missions, only that the Air Force's strengths lie in other areas.

As we have just discussed above, Air Force leadership of combat operations is more controversial. Doctrine indicates that the service that provides the preponderance of forces should take the lead in the

---

³¹ Interviews with Army officers responsible for command and control issues, Fort Leavenworth, Kan., November 2006.

³² Lt Col Howard D. Belote, USAF, *Once in a Blue Moon: Airmen in Theater Command—Lauris Norstad, Albrecht Kesselring, and Their Relevance to the Twenty-First Century Air Force*, Maxwell Air Force Base, Ala.: Air University Press, CADRE Paper No. 7, July 2000.

operation.³³ This standard is commonly used at the component level, but at the joint level, comparing numbers of personnel and platforms across forces operating in different mediums leads quickly to an "apples and oranges" problem. A more appropriate assessment would be based on the nature of the operation. To the extent that a campaign relies on forces from a particular medium, it would be most appropriate that the leadership of the operation have direct knowledge of operating in that medium. The evolving role of air power, and the uneven degree of acceptance of this role, will continue to spark debate. Nevertheless, a case can be made for the Air Force to lead combat operations of considerable size. One prominent retired Air Force officer suggests that the Air Force could lead operations up to a fight at the level of division/Marine expeditionary unit/Air Expeditionary Wing.³⁴ One could certainly make the case that it would have been appropriate to have an Air Force officer lead OAF or Operation Enduring Freedom, given the central role of air power in both conflicts.³⁵

To improve its capability to lead more types of JTFs, the Air Force must have a thorough understanding of the requirements of leading a JTF, and it must develop convincing responses to several challenges associated with the enterprise. This monograph seeks to deepen that understanding. The next chapter discusses how different members of the joint force think about what it takes to build JTF commanders and staffs.

---

³³ For example, the key doctrinal document concerning JTFs suggests that when designating functional component commanders, the joint force commander should appoint an officer from the service that contributes the most assets to be tasked and the command and control over them (JP 3-33, 2007, p. III-2). Air Force doctrine makes a similar suggestion for assigning the joint force air component commander, see Headquarters, U.S. Air Force, *Air Force Basic Doctrine: Air Force Doctrine Document 1,* Washington, D.C., November 17, 2003, pp. 64–65.

³⁴ Interview with Lt Gen (ret) Joseph Hurd, July 25, 2007.

³⁵ For an assessment of the role of air attacks in bringing the Serbian leader to capitulate in OAF, see Stephen T. Hosmer, *The Conflict over Kosovo: Why Milosevic Decided to Settle When He Did,* Santa Monica, Calif.: RAND Corporation, MR-1351-AF, 2001. For a review of air operations and their role in the defeat of the Taliban in Afghanistan, see Benjamin S. Lambeth, *Air Power Against Terror: America's Conduct of Operation Enduring Freedom,* Santa Monica, Calif.: RAND Corporation, MG-166-1-CENTAF, 2005.

CHAPTER THREE
Command Concepts

While considering how the Air Force would stand up JTF headquarters, it is worth considering how others approach the issue of command. Air Force officers in JTF leadership roles will need to understand how components provided by other services function in order to ensure productive collaboration. An appreciation for the different approaches of sister services and of other DoD entities also helps to highlight general factors that affect military command. The precise features of organizational charts are less important here than a broader perception of the perspectives of other elements of the joint force. This chapter discusses two "themes" that seem to recur in current American military command staffs. It then examines recent initiatives and general approaches relating to command in each of the services, within JFCOM, and in the Office of the Secretary of Defense. We refer to these initiatives and the thinking behind them as "command concepts."[1] As we will see, different elements of the U.S. defense establishment are pursuing a number of initiatives. Most of the developments in the services focus on the medium-based components that they provide to combatant commands.[2] Nevertheless, they have implications for JTF head-

---

[1] This is different from the type of "command concepts" discussed in Builder, Bankes, and Nordin, 1999. Builder, Bankes, and Nordin refer to commanders' conception of an operation, while we refer to the guiding principles and motivations behind various initiatives relating to command.

[2] That is, the Army and the Marine Corps field units that operate primarily from the ground, while the Navy operates from the sea, and the Air Force from the air.

quarters-level command as well. Lastly, this chapter considers some developments in the Air Force command community.

Themes

Employment and Management

There are different types of military staffs. Some staff officers, such as those on the Joint Staff, work at the national level. Others work at the departmental or service levels, supporting the Department of the Army, for example. Other "field" staffs assist military commanders who engage directly with U.S. security partners and prepare to fight U.S. adversaries.[3] We posit that field staffs must strike a balance between two types of work: employment and management. The boundary between the two is blurry, but we believe that it is still worthwhile to attempt to distinguish between them. When a staff plans and oversees the use of military means in an operation, it is engaged in employment. When a staff prepares military means for employment, it is providing management. For the Air Force, AOC personnel primarily work on employment, while Headquarters, Air Force, and major command (MAJCOM) staffs primarily fulfill management roles.[4]

For geographic combatant command, component-level staffs, employment work relates to efforts to plan and execute the actions of military forces. Management relates to their efforts to receive and support forces in the theater. A staff can perform both employment and management functions. Some of the command concepts we discuss

---

[3] For more on the different types of staffs, as well as a history of the military staff, see Lt Col J. D. Hittle, USMC, *The Military Staff: Its History and Development*, Harrisburg, Pa.: The Military Service Publishing Company, 1949. Robert Worley distinguishes between two separate chains of command, one that governs the use of force (operations) and one that governs the production of force (management). See D. Robert Worley, *Shaping U.S. Military Forces: Revolution or Relevance in a Post–Cold War World*, Westport, Conn.: Praeger Security International, 2006, pp. 3–4.

[4] This is a generalization that is ripe for contradiction. In particular, Air Force MAJCOM staffs associated with geographical combatant commands, such as PACAF and USAFE, do engage in a number of tasks that could be considered employment. We are grateful to Lt Gen Philp M. Breedlove and Maj Gen James P. Hunt for discussion on this point.

separate these functions by having different officers and suborganizations execute them, while others have the same staff work on both types of tasks.

Separating the two functions can allow a staff or portion of the staff to focus almost solely on military operations. Reducing the variety of demands on a staff can help it narrow its efforts and devote itself to employment without the distraction of management. On the other hand, in the absence of a contingency or other pressing operational assignments, much of the "action" on a staff often involves management. This can become a problem when it comes to assigning officer positions. Ambitious officers will seek posts that offer the opportunity to work on important tasks with high-ranking officers. Accepting a position at an operationally focused staff can mean accepting the risk that little of import will happen during one's tour. Thus, officers face something of a dilemma. They could take an employment job where nothing may happen but could be an important place to be in the event of a contingency, or they could take a management position that might feature regular interaction with high-ranking superiors and the chance to take part in an administrative initiative valued by the service. Officers in management positions could also be seconded to operational posts in the event of a contingency to help augment core staff, which means that there is even less incentive to seek an assignment on an employment staff.

This discussion so far relates primarily to the considerations of Air Force officers. For Army officers, work on an employment staff, such as a division staff, is perceived as helping to gain promotion. Air Force platforms fight as packages that are sized and configured to mission needs, not as squadrons or wings. It is the AOC that exercises command during a conflict. In contrast, divisions, corps, and theater armies traditionally fought as they are organized during steady-state operations. A division staff can work under a Combined Force Land Component Commander, but it can have command responsibility during operations. As we will see, Army commanders of field organizations fulfill employment functions as well as management ones. In contrast, Air Force squadron and wing commanders carry out management functions more than employment ones.

This issue is an important one for DoD to consider when building command elements for joint operations. Dedicated, "standing," headquarters that focus on force employment bring with them the advantages of readiness and competence. On the other hand, it can be difficult to recruit and maintain well-qualified personnel to work in standing employment staff organizations that do not have management functions. Giving headquarters administrative responsibilities means that officers with high potential can have more substantive work to do in the absence of a contingency, but it can dull the organization's operational edge.

Time and Function

We also found that staffs can be organized around time, function, or a combination of the two. Time-focused staffs tend to organize their activities around the daily "battle rhythm." For example, a "futures" cell considers overall strategy and longer-range plans. A near-term cell plans for the shorter term, and yet another cell focuses on overseeing current operations. A functionally based staff, on the other hand, focuses on the nature of tasks. It could feature cells that focus on intelligence, mobility, fires, maneuver, or other military tasks.

Generally speaking, the "closer" an employment staff is to tactical operations, the more it will tend to favor time over function as its central organizing imperative. Overseeing tasks that require close, constant, and quick collaboration between the control center and fielded forces indicates that time is a priority. Staffs that focus more on management and staffs that craft broader strategy are less time-sensitive and tend to be function based.

As we mentioned, time- and function-centered entities can be combined in a staff. A time-based staff might have an assessment cell that takes information from the current operations cell and feeds it to the future or near-term cells. A function-based staff might have a cell that provides current information to the rest of the organization.

If a staff oversees operations, the nature of the operations, particularly the medium in which the operations occur, can also affect whether the staff is time or function based. Directing air assets is not the same thing as directing ground or sea-based assets, leading each service-based component to emphasize different organizational constructs. Also, command at the component level, where it is necessary to closely monitor and direct fielded platforms and troops in the field, places a premium on time. Command staffs at the JTF headquarters level need to be sensitive to time, but they can afford to place more emphasis on military functions and the broader picture that emerges from individual-level encounters.

Army

The Army is accustomed to taking a leadership role in JTF headquarters. As we have shown earlier, Army units have served as the core of JTF headquarters in over half of the JTFs since 1990. The Army is currently restructuring the way it presents forces. In order to provide units that can be more readily tailored to the needs of a particular operation, the Army is deemphasizing the role of the division and emphasizing the "modular" brigade. In a way, this shift will actually make the Army look more like the Air Force. Air Force units that fly air-breathing platforms are organized into squadrons and wings, but squadron and wing commanders do not exercise control over their forces when they are employed in an operation. Instead, platforms from a variety of squadrons, wings, services, and even countries are bundled together into packages, and flights are controlled by the AOC. In contrast, Army units have traditionally operated as they are organized in peacetime. Brigades could be assigned to support different corps and other division headquarters, but for the most part, they were controlled by their "parent" division headquarters. The current approach aims to increase flexibility by making individual brigades more able to operate independently.

The Army also puts an emphasis on providing headquarters elements for JTFs. In fact, the *primary* role of the Army's corps headquarters is to serve as the core of a JTF headquarters.[5] Theater army and

---

[5] U.S. Army Training and Doctrine Command, *The Army Modular Force,* Fort Leavenworth, Kan.: Combined Arms Center, FMI 3-0.1, November 17, 2006, pp. 1–8. This is a

division headquarters are also charged with being prepared to assume the role of a JTF headquarters with appropriate augmentation, although this is not considered to be their main focus.

All three levels of organization, the theater army, corps, and division, are also to be capable of serving as core elements of joint force land component commands (JFLCCs). They are also designed to be able to provide administrative support to all Army forces (ARFOR) in theater. We should also note here that for JTFs led by two-star generals and above, the Army's divisions, corps, and "theater armies" (described below) provide three separate levels of options for command, as opposed to the Air Force's single option—the Numbered Air Force (NAF).[6] Having different echelons of command makes it relatively simple for the Army to provide both JTF and component-level headquarters. If a theater army is named to lead a JTF headquarters, a corps or division staff can staff the bulk of the JFLCC. If a corps staff is called upon for the JTF headquarters role, a division staff can serve as the JFLCC. In contrast, since the Air Force has only NAF and MAJCOM above the wing level in a theater, it is unclear how the Air Force would fill both the JTF headquarters and Combined Force Air Component Command (CFACC) roles simultaneously.

The Army model for JTF operations is for the Army-supplied JTF headquarters to deploy to a theater. Air Force command elements, however, do not usually deploy. With the exception of the U.S. Air Force Special Operations Command's Warfighting Headquarters and Ninth Air Force's current operation in the Middle East, NAFs and their associated AOCs do not deploy for operations. For the Air Force, the platforms deploy, and there is less of a perceived need for the headquarters element to be "near" the area of operations.

The Army's leadership has put a good deal of thought into how it will integrate SJFHQ (CE) and other elements into the Army's command structures to form JTF headquarters. It has thought about how

---

working document and is subject to change, but it does accurately reflect Army thinking about its command and control structures as of the writing of this monograph.

[6] The Air Force also has MAJCOMs that oversee NAFs, but MAJCOMs do not traditionally command and control fielded forces.

to integrate joint augmentees into its existing organizations, down to the level of detail of individual slots, as is depicted in Figure 3.1. In December 2004, the U.S. Army's Training and Doctrine Command and JFCOM formed a "tiger team" to discuss how to transform Army headquarters into JTF headquarters.[7] To some extent, one could say that the Army has a head start on writing the book about how to form JTF headquarters.

A theater army headquarters serves as the Army service component command for each of the geographical combatant commands. In this capacity, it also provides administrative control of Army forces in its AOR. A theater army is also prepared to provide operational

Figure 3.1
Forming the Joint Planning Group: Example

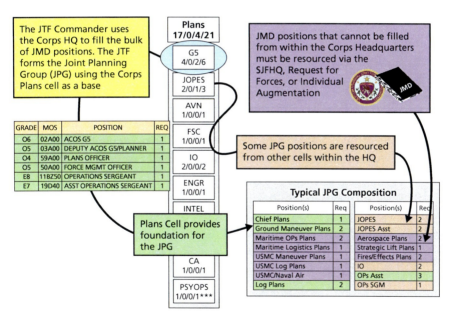

SOURCE: "Modularity Echelon Brief," July 23, 2007.
RAND MG777-3.1

---

[7] Combined Arms Doctrine Directorate, "Army Organizations as Joint Capable Headquarters," briefing, August 13, 2007b, p. 27.

control of forces and can serve as the core of either a JTF headquarters or a JFLCC. A theater army has two separate command entities, a main command post (MCP) and an operational command post (OCP). The MCP, which consists of 588 people, allows for command of administrative functions and for Army support to other services. The OCP, which is deployable, allows the commander to control "operational-level" forces. The OCP has 481 positions.[8] Both the MCP and the OCP are organized along functional lines, although they break away slightly from the traditional J-code structure, with directorates for movement and maneuver; fire support; command and control; operational sustainment; operational protection; and intelligence, surveillance, and reconnaissance (ISR).[9] The new structure is directed from headquarters, though; and initial reports from the field indicate that some units still prefer to keep elements of the J-code structure. A staff member of Third Army has reported that it has a command group, an operational sustainment group (G-1, G-4, and medical staff), an intelligence section, an operational maneuver section (G-3, 5, 7; plans, current operations, future operations, and fires and effects), a communications group (G-6), and an operational protection group (military police, air defense, and chemical protection).[10] Even if other field elements do not fully implement the new structure, it does appear that there is movement away from traditional staff organization.

As we mentioned above, the corps headquarters' primary role is to serve as the core of a JTF headquarters. It can also serve as the main element of a JFLCC headquarters, or as an ARFOR headquarters to support Army forces within a geographic combatant command by providing administrative support. A corps headquarters has a mobile command group to allow the commander to gather information and pro-

---

[8] Personnel numbers for MCP and OCP are from Combined Arms Doctrine Directorate, *Command and Control in the Modular Force: Echelons and the Road to Modularity,* July 3, 2007a, p. 26.

[9] U.S. Army Training and Doctrine Command, 2006, Chapter 3. At present, MCP and OCP are organized along J-codes such as J-1 (personnel), J-2 (intelligence), J-3 (operations), J-4 (logistics), and J-5 (plans).

[10] Email to one of the authors, May 17, 2007.

vide guidance in the field. It also has an MCP and a tactical command post (TAC CP). The Army is currently planning for the MCP to have 525 positions and for the TAC CP to have 77.[11] TAC CP maintains communications with higher Army headquarters for matters relating to training, organizing, and equipping the unit, while the MCP oversees current operations, conducts future planning, and coordinates logistics support.

The corps TAC CP is functionally organized, and like the theater-army MCP, it breaks with the traditional J-codes, with cells for command, control, communications, and computers (C4); fire support; sustainment; force protection; ISR; and movement and maneuver. The MCP has similar warfighting functions, but places them within a time-based organization. It has a current operations cell, a future operations integrating cell, and a plans cell.[12]

The division, the "Army's primary warfighting headquarters," seeks to aggregate tactical actions to meet operational-level objectives.[13] Like the corps, the division has a mobile command group, an MCP, and a TAC CP. The TAC CP is designed to control narrowly defined operations, such as a river crossing. The MCP works mainly through three time-based cells, which focus on plans, current operations, and future operations. Within each of these time-based cells are elements that deal with different warfighting functions, such as fires, air defense, and intelligence. There are also cells devoted to warfighting functions, some of which duplicate the elements within the time-based cells. In the event that a division commander is asked to form a JTF headquarters, he or she can retain this organization or convert it back to a standard J-staff.[14] The TAC CP has 72 billets, while the MCP has 450. At present, the Army is looking to reduce the number of positions in its divisions by 5 percent.[15]

---

[11] Combined Arms Doctrine Directorate, 2007a, p. 44.

[12] U.S. Army Training and Doctrine Command, 2006, Chapter 4.

[13] U.S. Army Training and Doctrine Command, 2006, p. 5-1.

[14] U.S. Army Training and Doctrine Command, 2006, p. 5-24.

[15] Combined Arms Doctrine Directorate, 2007a, pp. 40, 42.

From the Army's plans for its command staffs, we can see the interplay of time-based and function-based organizations. At each echelon, the Army is moving away from the traditional J-codes to a different set of military functions. At the corps and division levels, these functions are subsumed within time-based cells, as Figure 3.2 shows.

Army command staffs are also designed to deal with both management and operational responsibilities. At the theater-army level, it appears that the MCP deals with most management tasks, leaving the OCP to handle operations. For the corps, the TAC CP has responsibility for communicating with the institutional Army, but the OCP handles resource management, provost marshal, and some other

Figure 3.2
Army Combines Function-Based and Time-Based Elements

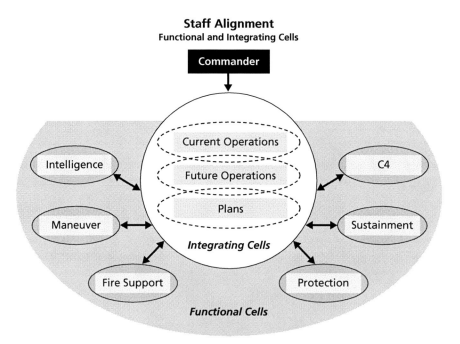

SOURCE: Combined Arms Doctrine Directorate, 2007a, p. 49.
NOTES: Not all staff sections and functional cells are permanently represented in all of the integrating cells. They do, however, provide representation as required.

management tasks. The Army has taken subordinate units away from its corps headquarters as well as the units' responsibility to provide Title 10 support to division headquarters in order to allow the units to focus on preparations to assume the JTF, JFLCC, or ARFOR roles.[16] Division MCP also deals with both management and operational tasks.

The Army's preparations for its headquarters go beyond writing papers about organizational design. The Army's Battle Command Training Program (BCTP) trains personnel to work on brigade, division, corps, theater-army service component command, JFLCC, and JTF staffs.[17] With a staff of over 600, BCTP puts units through a seminar and an exercise before they deploy for operations.

Navy

The Navy is in the midst of implementing its own command initiative, the Maritime Headquarters with Maritime Operations Center (MHQ with MOC). After the experiences of operations in Iraq, Afghanistan, and elsewhere, the Navy is concerned about its ability to command in a changing environment, obtain situational awareness, and work with other members of the joint force. It is interesting that the Navy would like to emulate some of the successes of the Air Force's AOC development efforts. The Navy seeks to standardize its staffs, their functions, and the processes they follow. It also seeks to increase the Navy's ability to command and plan at the operational level of war.[18]

This is an ambitious step. At 300–500 people, numbered fleet staffs tend to be smaller than their Air Force and Army counterparts. The requirement for officers to serve on ships means that there is less manpower available for Navy staffs. These smaller staffs have been able

---

[16] Combined Arms Doctrine Directorate, 2007a, p. 33. Title 10 functions refer to service training, organizing, and equipping responsibilities.

[17] Battle Command Training Program, "Battle Command Training Program: Commander's Overview Briefing," November 16, 2006, p. 7.

[18] U.S. Navy, Second Fleet, *Maritime Headquarters with Maritime Operations Center Concept of Operations (MHQ with MOC CONOPS)*, Final Draft Version 2.4, Norfolk, Va., October 31, 2006, p. 10.

to fulfill the needs of traditional Navy "watches," but the shift from a watch that focuses on tracking ships to an operations center that senses, plans, and commands forces to achieve operation-level effects is considerable.[19]

As part of this shift, the Navy MHQ with MOC concept of operations (CONOPS) calls for several measures to train and educate naval officers in operational art. It recommends the establishment of a senior mentor program, increasing attendance in the naval operational planner's course, creating a Web-based course for MOC staff, and building a tracking mechanism to make it easy to locate reserve officers with skills amenable to MOC duty.[20] There are also proposals for a MOC operator's course of five to six weeks, and a one-week Joint Force Maritime Component Command/Combined Force Maritime Component Command (JFMCC/CFMCC) course.[21]

Like Army theater armies, corps, and division headquarters, the Navy's numbered fleet headquarters are supposed to be capable of forming the core of JTF headquarters. In fact, as of the writing of this monograph, Second Fleet is in the process of JTF headquarters certification with JFCOM. The first two MHQs with MOCs were established at Second Fleet, based in Norfolk, Virginia, and Fifth Fleet, based in Manama, Bahrain. MHQs with MOCs will eventually be created for each of the combatant commands.

MHQs with MOCs work on both operations and administrative matters. At Second Fleet, roughly one-third of the MHQ staff works on fleet management, which includes efforts to prepare forces for employment by combatant commanders, to maintain reserve forces, to develop doctrine and tactics, techniques, and procedures, and to manage program and budget requests and other matters.[22] Another third of the staff splits its time between fleet management and opera-

---

[19] For more information on the nature of the watch, see ADM James Stavridis, USN, and CAPT Robert Girrier, USN, *Watch Officer's Guide: A Handbook for All Deck Watch Officers*, 15th ed., Annapolis, Md.: Naval Institute Press, 2007.

[20] *MHQ with MOC CONOPS*, 2006, p. 60.

[21] U.S. Navy, Second Fleet, untitled MHQ with MOC brief, not dated, p. 34.

[22] *MHQ with MOC CONOPS*, 2006, p. 13.

tions, and the remaining third focuses on operations.[23] In contrast, the Fifth Fleet tends to separate fleet management from operations. MOC staff can be called upon to work fleet management issues for large or urgent efforts, or if they possess particularly useful skills. But for the most part, they focus on operational matters.[24]

Both Second Fleet and Fifth Fleet have moved away from the traditional Napoleonic, N-code organization, and each addresses the issue of time versus operations. Figure 3.3 shows the design of Fifth Fleet. The Fifth Fleet MHQ with MOC construct emphasizes time, with

**Figure 3.3
Fifth Fleet Organization**

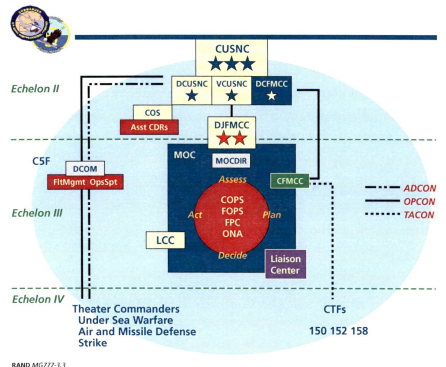

---

[23] Interview with Second Fleet Staff, April 5, 2007.

[24] Email from Fifth Fleet Staff, May 16, 2007.

centers for current operations (COPS), future operations (FOPS), future plans (FPC), and operational net assessment (ONA).[25] The MOC director (MOCDIR) oversees these cells. Functional specialists with operations, intelligence, and logistics experience have been placed in each of these centers in interdisciplinary teams. Second Fleet is organized a bit differently. In addition to the four time-based centers, the fleet has functional organizations for logistics and intelligence.[26]

Naval operations create requirements that are different from air operations. Air platforms generally carry out one to three missions in a discrete 24-hour period, and then they are available again the next day. Maritime platforms can be on station for months at a time and engage in longer-duration missions than do air platforms. As a result, naval planners plan for more than one mission at a time. There is no "maritime tasking order" akin to the air tasking order (ATO), although there are some who argue that instituting one would help naval components focus on capabilities instead of platforms.[27]

One more aspect of the MHQ with MOC concept deserves mention here. Naval components often consider themselves more as Navy commands than as maritime or coalition components. One vice admiral describes this as being "inherently joint from a service perspective."[28] We interpret this to mean that he uses primarily Navy forces (NAVFOR) to produce effects for the joint combatant commander. For example, as currently organized, Fifth Fleet is the Navy component to the Central Command (CENTCOM), and it is led by a three-star flag officer. There is a position for a two-star deputy who can serve as the joint force, maritime component commander. There is also a position for a one-star deputy who can serve as a coalition-force maritime component commander. Thus, the coalition and joint roles are separate from one another and from the naval component role.

---

[25] "NAVCENT MHQ w/MOC," briefing, March 27, 2007.

[26] Interview with Second Fleet staff, April 5, 2007.

[27] Interview with Fifth Fleet staff, April 15, 2007.

[28] Email from maritime component commander, April 29, 2007.

Marine Corps

In some ways, the Marine Corps is the epitome of combined arms operations. Marine expeditionary forces (MEFs) work across the domains of land, air, and sea. Because of its multidimensional nature, however, the Marine Corps presents challenges when integrated with other elements of the U.S. military. Marines are hesitant to split their forces between the air and land components. Unlike the Air Force, Army, or Navy, the Marine Corps as a service does not fit neatly into the medium-based component model that DoD has favored in recent operations.

The Marine Air-Ground Task Force, as its name implies, combines air and ground assets into a single unit. Marine pilots train and work regularly with the same ground units, building relationships and common views on how to fight. Marine aviators have tended to focus on close air support instead of deeper attacks on adversary infrastructure and other "strategic" targets. During Operation Iraqi Freedom, the Air Force and the Marine Corps agreed to include Marine aircraft in the ATO and to place them under the control of the combined air operations center, on the condition that they be sent back out (when possible) to support Marine ground forces.[29] The agreement worked fairly well.[30] Since the overthrow of Saddam Hussein's regime, however, Marines operating in western Iraq have not put their aircraft under the operational control of the Joint Force Air Component Command (JFACC), causing much consternation within the Air Force. The tension between USMC's desire to operate as a single unit and the Air Force's desire to consolidate all air assets under one joint commander remains unresolved.

The MEF, led by a three-star general officer, is the Marine unit deemed capable of forming the core of a JTF headquarters. MEF staffs vary from 180 to 300 people. MEFs consist of a command element, a ground combat element, an air combat element, and a combat logis-

---

[29] Then–Lt Gen T. Michael Moseley allowed the Marine Corps to direct its own aircraft in the area where I MEF operated, under the condition that it remained on the ATO and under the control of CFACC. I am grateful to Benjamin S. Lambeth for discussion on this point.

[30] Rebecca Grant, "Marine Air in the Mainstream," *Air Force Magazine*, Vol. 87, No. 6, June 2004.

tics element. MEF command elements are time based, featuring future operations, plans, and current operations cells.[31] They also combine management and operations.

Like the Army, the Marine Corps runs a program to exercise its staffs. The Marine Air-Ground Task Force Staff Training Program (MSTP) focuses on training MEF command element staff. It is charged with completing a full training cycle with each of the three MEF staffs at least once every two years, although recent operational demands have pushed the staffs to a higher operational tempo.[32] MSTP normally sends its trainers and observer/controllers to unit home stations to conduct training, which includes a command post exercise.

Joint Force Command

As a combatant command that places much of its effort on developing future capabilities, JFCOM plays a role in a number of initiatives that deal with command and control. Above we discussed the Standing Joint Force Headquarters (Core Element) initiative, which seeks to supply an SJFHQ in each combatant command, along with two more at JFCOM. SJFHQ (CE) consists of a command group and four functionally based teams: information superiority, plans, knowledge management, and operations.[33] Implementation of the initiative has been uneven. The main difficulty has been securing the necessary manpower to fill the 57 positions of the command element. There are also problems with locating the funding necessary to train personnel for SJFHQ.[34]

Each combatant command has used the SJFHQ (CE) in a different way. For example, the United States European Command (USEUCOM or EUCOM) has created a EUCOM Plans and Operations

---

[31] Interview with Marine Corps Concept Development Command staff, June 1, 2007.

[32] Email from MSTP staff, September 19, 2006.

[33] U.S. Joint Force Command, 2007.

[34] U.S. European Command, "USEUCOM Training Transformation in Support of C2 Transformation," briefing, September 2003, p. 6.

Center (EPOC).[35] Contrary to most expectations of how the SJFHQ (CE) will work, the EPOC plans to deploy only around 20 of its staff to JTF headquarters. U.S. Pacific Command (PACOM) has 19 people in its J-7 staff, and it has spread the remaining posts across its staff.[36] The 2006 QDR initiative, which calls upon service components to be capable of forming the core of JTF headquarters, could be viewed as an attempt to redress the deficiencies in the SJFHQ (CE).

JFCOM is also heavily involved in the Command and Control Capabilities Portfolio Management initiative and the Command and Control Capabilities Integration Board. Both initiatives involve overseeing the development and acquisition of command and control systems. The services have expressed alarm that the initiatives might usurp their traditional Title 10 roles. The two initiatives are also related to efforts to build service capabilities to provide the core of JTF headquarters. JFCOM has established "templates" for JTF headquarters that lay out suggested organizational schemes and the necessary systems to operate a JTF headquarters. There are templates for major operations and campaigns, stability operations, and disaster relief.

JFCOM J-7 is developing an exercise program to help train and certify service components as being capable of leading JTFs. As we mentioned above, the Navy's Second Fleet is currently undergoing certification, which JFCOM views as a "proof of concept" for how to prepare JTF headquarters. JTF certification, however, is still a work in progress. In addition to JFCOM's certification initiative, some of the combatant commands have programs of their own. For example, EUCOM has its own program, which includes its own set of mission-essential tasks that JTF headquarters need to master to be considered ready. EUCOM has certified both Sixth Fleet and Third Air Force as JTF capable.

---

[35] Hugh C. McBride, "New Plans and Operations Center Exemplifies EUCOM Transformation," *American Forces Press Service*, October 30, 2003.

[36] Interview with JFCOM staff, April 5, 2007.

36 What It Takes: Air Force Command of Joint Operations

The SJFHQ office at JFCOM has also developed a concept of operations for setting up standing JTF headquarters.[37] The first draft of the document received a difficult reception from the services. Air Force officers viewed the document as being ground centric and not suited to Air Force capabilities and perspectives. The Army had its own reasons for opposing the document, with some disputing JFCOM's knowledge in the area of command and control. Instead of being told what systems to procure, Army officers argued that they should be issued standards of performance to meet in whatever manner they decide is best.[38]

There are a number of other initiatives, programs, and concepts dealing with command and control at JFCOM and elsewhere. For example, JFCOM has been involved with efforts to create a joint force support component command, a logistics command that would have more or less equal standing along with medium-based commands such as JFACC and JFLCC. There have also been calls for a joint information commander, and under Secretary of Defense Donald Rumsfeld, DoD created the Defense Joint Intelligence Operations Center (DJIOC). Based in Washington, D.C., DJIOC represents an attempt to couple intelligence with operations.[39] It is related to the establishment of the Joint Functional Component Command for ISR at the Strategic Command.

Air Force

The component–Numbered Air Force (C-NAF) initiative seeks to standardize Air Force contributions to combatant commands. It attempts to direct as many administrative tasks as possible to Air Force MAJCOMs, allowing C-NAF staff to focus on planning and operations. The Air Force argues that a focus on operations will help foster more

---

[37] Headquarters, Standing Joint Force, "Improving Readiness for Joint Task Force Headquarters: Concept of Operations," unpublished JFCOM research, July 13, 2007.

[38] Interview with Army officers at Fort Leavenworth, Kan., November 14, 2006.

[39] Defense Joint Intelligence Operations Center, "The Defense Joint Intelligence Operations Center: Overview," briefing, September 13, 2006.

knowledge of the region, build a more-coherent team, and result in a readier, more-effective air component. Air Force documents also indicate that C-NAFs should be capable of forming the core of JTF headquarters if called upon to do so.[40]

The C-NAF structure, shown in Figure 3.4, is made up of AOC and an Air Force Forces (AFFOR) staff. AOC is primarily time based, with divisions for strategy, plans, and operations. Long-term objectives are set in the strategy division. The planning division translates these

Figure 3.4
C-NAF Internal Structure

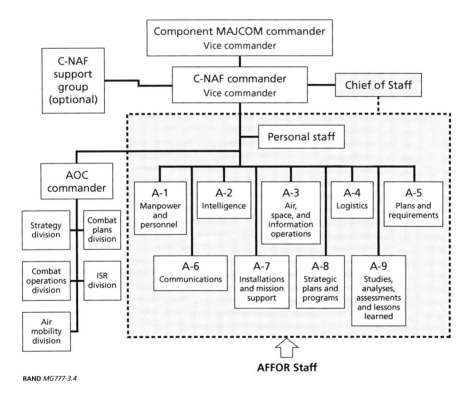

[40] Headquarters, U.S. Air Force, *Implementation of the Chief of Staff of the Air Force Direction to Establish an Air Force Component Organization*, Program Action Directive 06-09, September 15, 2006b, pp. A-3, A-4; and Headquarters, U.S. Air Force, "Air Force Forces Command and Control Enabling Concept," Washington, D.C., May 25, 2006a, pp. 1, 6, 7.

notions into a master air-attack plan and an ATO, which is then executed by the operations division. The strategy division then assesses the results of operations and, with input from leadership and others, makes necessary adjustments. The AOC also has two functional divisions that focus on ISR and mobility, but the focus of the AOC staff is on time, not function.

In contrast to the time-based AOC, the AFFOR staff is based on the functional J-code model. The AFFOR staff tends to work on management tasks, while the AOC works on employment. The AFFOR staff is more employment oriented than are the staffs at MAJCOM headquarters or the Pentagon-based Air Staff, but it is less focused on the execution of current operations than is the AOC. The AFFOR staff assists in force deployment, force bed down, and sustainment; it conducts adaptive planning and assists the commander of Air Force Forces (COMAFFOR) in theater engagement.[41] To the greatest extent possible, AFFOR and AOC personnel are not dual-hatted, which helps enforce a separation between management and employment within C-NAF.

The Air Force expects C-NAF headquarters, and more specifically AFFOR staffs, to be capable of forming the core of a JTF headquarters.[42] As discussed later in this monograph, while the AFFOR staff will likely form the core of a JTF headquarters staff, the AFFOR staff is not optimized in day-to-day operations to provide command and control of joint operations.

In addition to C-NAF, the Air Force is sponsoring other command-related initiatives. As mentioned above, one NAF headquarters has actually obtained certification that it is capable of forming a JTF headquarters core. In the summer of 2007, Third Air Force participated in Flexible Leader, a EUCOM exercise that had the C-NAF staff play the role of a JTF headquarters core.[43] The exercise simulated an

---

[41] Headquarters, U.S. Air Force, 2006a, p. 18.

[42] Headquarters, U.S. Air Force, *Operations and Organization,* Air Force Doctrine Document 2, Washington, D.C., April 3, 2007b, p. 113.

[43] Interview with Phillip M. Romanowicz, Chief, Exercises and Training Analysis, 3 AF/A9X, June 4, 2007.

earthquake and resulting humanitarian disaster in Turkey. This scenario is reminiscent of Third Air Force's role in JTF–Atlas Response, which is discussed later in the monograph. As a result of its work in Flexible Leader, EUCOM certified Third Air Force as being ready to undertake the JTF headquarters role.

Other command-related initiatives are less immediately relevant to the issue of standing up a JTF headquarters, but they do play an important role in shaping USAF command capabilities. For example, the Air Force is in the midst of creating an operations support facility (OSF). The OSF, based at Langley Air Force Base, is tasked with providing data backup and continuity of operations for AOCs around the globe. It is also intended to assist with training, exercise support, and experimentation for AOC and AFFOR staffs. In addition, the Air Force hopes to minimize the footprint of AOCs and reduce their need for augmentation by having them "reach back" to other AOCs or by having some of their work performed at OSF.[44]

The Air Force is also building an "ISR command," which appears to be more of a field operating agency than an operational headquarters. The Air Force created an A-2 directorate in its headquarters and consolidated its different intelligence organizations into a single command.[45] One objective of the move is to orient Air Force intelligence around capabilities instead of dislocated programs.[46] The new command focuses on developing new capabilities rather than on directing or overseeing operations.[47] It also seeks to develop a cadre of Air Force

---

[44] Air Combat Command, "Air Force Forces Component Numbered Air Force Operations Support Facility Functional Concept," November 8, 2006.

[45] Headquarters, U.S. Air Force, "USAF Intelligence Way Ahead," briefing, Washington, D.C., January 16, 2007a.

[46] Gayle S. Putrich, "USAF Reorganizing Intelligence Command," *Defense News*, January 30, 2007.

[47] Robert K. Ackerman, "New Flight Plan for Air Force Intelligence," *Signal Magazine*, March 2007.

general officers who can be considered for joint positions, such as the J-2 on a combatant command staff.[48]

The scope of these initiatives is quite broad, particularly when one considers that the Air Force has recently cut personnel and is considering other cuts in an environment of fiscal austerity.[49] The Air Force currently has trouble filling the positions required to fully staff its AOCs.[50] It is unclear to what extent it will implement the initiatives discussed above. It has been doing more work with less manpower for some time.

Adding the capability of forming the core of JTF headquarters on top of existing initiatives will require additional resources, in terms of money, personnel, time, and institutional emphasis. This monograph serves as a preliminary attempt to scope out these costs, so that the Air Force can decide whether to and how it might meet them.

---

[48] SSgt C. Todd Lopez, USAF, "Changes Planned for ISR Community," *Air Force Print News*, January 30, 2007.

[49] Vago Muradian, "USAF Struggles with Budget Shortfall," *Defense News*, August 20, 2007, p. 4.

[50] Headquarters, U.S. Air Force, *Air and Space Operations Center Crew Roadmap*, Washington, D.C., September 20, 2006c.

CHAPTER FOUR
Lessons from Past JTFs

Examining previous examples of JTF headquarters can be a useful way to understand what will be required to prepare future ones. Among the variety of types of military operations, we chose to focus on humanitarian and combat operations. Within each of these categories, we chose one JTF for which an Air Force organization formed the core of the headquarters and another for which an entity from another service filled that role (see Table 4.1). In each instance, we examined how the unit prepared, how its staff was composed, and what particular command-related challenges emerged during the operation. For the JTFs in which an Air Force unit did not form the core of the headquarters, we ask how the operation might have differed if Air Force officers had led and provided the core of headquarters staff.

Above we observed that command at the operational level involves translating strategic guidance into relatively discrete goals and an operational plan. The commander then sets priorities and allocates the means at his or her disposal to meet these goals. In each example below, we examine how the JTF commander and staff performed these tasks.

Table 4.1
Selected JTFs

	Air Force–Led	Other Service–Led
Humanitarian	Atlas Response	CSF-536 (USMC)
Combat	Southwest Asia	Noble Anvil (Navy)

JTF–Atlas Response: The Benefits of Preparation and Presence

In early 2000, southern Africa experienced heavier than normal seasonal rains. Then two separate storms, Cyclone Connie and Cyclone Eline, hit the region within 17 days of each another, causing flooding in Botswana, South Africa, Mozambique, Zambia, and Zimbabwe. The weather contributed to the deaths of around 400 people and affected the lives of more than two million.[1] Headquarters, Third Air Force, under the leadership of then–Maj Gen Joseph Wehrle, was tasked with forming the core of JTF–Silent Promise, which was subsequently renamed JTF-AR. This JTF, which focused most of its efforts on Mozambique, sought to coordinate and synchronize disaster assistance efforts, conduct search and rescue, distribute relief supplies, and provide aerial assessment of conditions on the ground. It was the first major deployment of U.S. military forces in Africa since the completion of Operation Restore Hope in Somalia in 1993.

Some members of the Third Air Force staff were accustomed to playing a leading role in providing humanitarian support to populations in need. In March 1999, the headquarters, which was led at the time by Maj Gen William Hinton, led JTF–Shining Hope to assist Albanian Kosovars displaced by Serbian ethnic cleansing.[2] After taking command of Third Air Force, General Wehrle used Exercise TrailBlazer 2000 and another exercise, named Dust Devil, to learn more about how to form the core of a JTF, how to manage a JTF, and how to work with international organizations (IOs) and nongovernmental organizations (NGOs). He proposed to his superiors at U.S. Air Forces Europe that Third Air Force would focus its training and planning efforts on humanitarian support, while Sixteenth Air Force would

---

[1] Maj Gen Joseph Wehrle, "Joint Task Force ATLAS RESPONSE: Commander's Perspective," briefing, not dated; and Mike Cohen, "Mozambique-Floods," *Portsmouth Herald*, March 7, 2000.

[2] Gen John P. Jumper, USAF, "Rapidly Deploying Aerospace Power: Lessons from Allied Force," *Aerospace Power Journal*, Winter 1999.

focus on combat operations in the European theater. They accepted.³ He also had Third Air Force staff learn more about Africa, where they would likely deploy, and plan for a range of possible contingencies, focusing on sub-Saharan Africa.⁴ These efforts led General Wehrle to the idea that in humanitarian operations, forces work most effectively when they seek to "fill in the gaps" in the operations of other groups by providing unique capabilities. He also took away the lesson that it is desirable to avoid the appearance that U.S. forces can do the job by themselves, which can alienate officials from other agencies, governments, and public opinion.⁵

In the midst of cyclone season, with reports of flooding hitting the international media, on February 18, 2000, EUCOM ordered Third Air Force to form a humanitarian assistance survey team to travel to southern Africa to assess the situation. Third Air Force's deliberate preparations to command joint humanitarian operations likely contributed to EUCOM's decision to turn to Third Air Force. On February 22, Cyclone Eline hit Mozambique, leading to renewed flooding. On March 1, U.S. President Bill Clinton announced that the United States would send aid to Mozambique, and on the same day, EUCOM issued a warning order naming General Wehrle as the JTF commander.⁶ The combatant command issued an execute order on March 4, and General Wehrle arrived at Air Force Base Hoedspruit, South Africa, on March 6.⁷ The execution order directed General Wehrle to

> [C]onduct military operations (in support of) the lead federal agency, the Department of State, to provide support to

---

3   Author interview with Lt Gen (ret) Joseph Wehrle, July 2, 2007.

4   Robert Sligh, *ATLAS RESPONSE: Official History of Operation ATLAS RESPONSE*, Ramstein Air Base, Germany: Third Air Force, 2000.

5   Wehrle, not dated; and Sligh, 2000.

6   A *warning order* is a preliminary statement about an expected upcoming operation that describes a situation, allocates forces, and sets command relationships. For more information, see JP 1-02, 2007, p. 580.

7   An *execute order* implements a decision by the president to initiate a military operation. For more information, see JP 1-02, 2007, p. 191.

HUMANITARIAN ASSISTANCE (HA)/DISASTER RELIEF (DR) operations in Southern Africa, to include Mozambique and neighboring states, as required to relieve suffering and prevent further loss of life. Mission objective is to provide immediate lifesaving and other humanitarian support until the situation allows for transition of these responsibilities to host nation(s) and international relief organizations.[8]

Most JTF aerial platforms were based out of Hoedspruit, and most of the JTF headquarters staff worked from there, but General Wehrle and a small staff contingent worked from Maputo, Mozambique. This JTF also established civil-military operations centers in Maputo and Beira, Mozambique, and a joint special operations task force (JSOTF) headquarters in Beira. General Wehrle placed himself and other JTF command elements in Mozambique to help coordinate with IOs and NGOs on the ground there and to symbolize U.S. commitment to help the country. After the operation, General Wehrle reported that having a physical presence forward in Maputo was essential to developing the relationships necessary to allow the JTF to accomplish its mission.[9]

The forces available to the JTF were mostly provided by the Air Force. Of the 800 personnel who served in the operation, around 730 were from the Air Force. General Wehrle made a request for Army helicopters, but the Joint Chiefs of Staff declined to send them.[10] Navy and Marine units were either too far away or unavailable to join the operation. The JTF was able to obtain some capabilities from the special operations community, most of whom constituted the JSOTF in Beira. The JSOTF also served as a base for three HH-60G helicopters and a tanker airlift control element.

General Wehrle set four basic tasks for the JTF: coordination and synchronization, search and rescue, relief supply distribution, and

---

[8] BCC [blind carbon copy] to 3AF [Third Air Force]/CCEA, "R031835Z Mar 00 EXORD for Operation ATLAS RESPONSE," email, March 2000.

[9] Interview with General Wehrle.

[10] Interview with General Wehrle.

aerial assessment.[11] The headquarters was predominantly sourced with Air Force personnel, although it did have an Army deputy. Figure 4.1 shows that 128 of 147 headquarters staff slots, or 87 percent, were filled by Air Force personnel. The lack of time between JTF activation and the beginning of the operation helps explain the makeup of the headquarters. There is no indication that lack of representation from other services hindered the effectiveness of the operation.

The JTF headquarters faced significant constraints in its planning and deployment. After years of international isolation due to apartheid, the South African military was not used to working with foreign militaries. The South African government was also wary of any action that might appear to infringe upon its sovereignty. The British government received verbal permission from South Africa to contribute to the relief effort, but the crew members of a British Tristar were arrested when they landed at Hoedspruit because the South Africans had not yet issued formal written permission for them to land. Members of the humanitarian assistance survey team also angered South African authorities by not having visas and for saying that they were in the

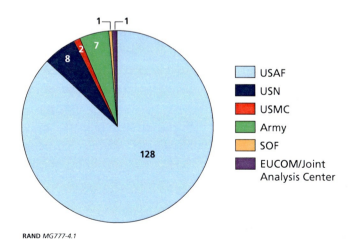

Figure 4.1
Composition of JTF–Atlas Response HQ, 147 Total Personnel

RAND MG777-4.1

---

[11] Wehrle, not dated.

country to conduct "flood relief," which touched upon South African sensitivities.[12]

The JTF headquarters also had to find ways to work with other militaries, IOs, and NGOs without ruffling feathers. For example, General Wehrle had wanted to synchronize aid efforts through the use of a process using an air coordination order, which was more or less akin to what Air Force officers refer to as an air tasking order, a plan that allocates and schedules available air platforms. Upon landing in South Africa, however, he learned that the other countries and organizations had already instituted a less efficient process that involved a daily meeting to allocate missions. To foster a spirit of collaboration, General Wehrle integrated U.S. efforts into the existing process rather than force the U.S. process onto others.[13]

In less than a month, JTF-AR delivered more than 700 short tons of cargo on behalf of international and nongovernmental aid organizations and transported aid workers and others around the area of operations. It also collected imagery from C-130s using a new system called Keen Sage and "lieutenants on ropes" using cameras to assess the state of local infrastructure on the ground and to locate concentrations of displaced people.[14] Politically, JTF-AR presented the United States as a caring and effective provider of assistance to those in need, while also demonstrating that U.S. forces can work well with interagency and international partners. It also showed that the Air Force can form the core of a successful JTF headquarters in a humanitarian operation. The experience indicates that efforts to prepare units through education and training have real payoffs. Early involvement and anticipation of their mission helped Third Air Force personnel accomplish this mission. Sensitivity to the needs of other actors also helped General Wehrle and his staff fit in with representatives of local governments, NGOs,

---

[12] Sligh, 2000, pp. 11, 14.

[13] Col S. Taco Gilbert, "DIRMOBFOR JTF ATLAS RESPONSE After Action Report," February 28, 2000; and Sligh, 2000, pp. 24–26.

[14] *Keen Sage* is a surveillance device that includes three sophisticated cameras, a daylight television camera, a 955mm fixed-focal-length zoom, and an infrared camera. "Lieutenants on ropes" are officers with digital cameras standing by the open doors of cargo helicopters.

and other militaries. JTF-AR illustrates the importance of political factors in humanitarian operations. It also shows that by preparing for this type of mission in advance, Headquarters, Third Air Force put itself in a position to lead the JTF once national authorities decided that one was needed.

JTF–Unified Assistance (CSF-536): Mixed Modes of Control

The Sumatra-Andaman earthquake that struck on December 26, 2004, triggered the most destructive tsunami in recorded history. At least 227,898 people were killed and more than one million were displaced by the 9.0 magnitude earthquake and tsunami.[15] The following day, PACOM headquarters, anticipating U.S. involvement, issued execute order 271009Z, which stated that, when tasked, PACOM would "assist in rapidly reducing loss of life, mitigate suffering, and reduce the scope of the disaster."[16] PACOM issued planning order 271115Z to assess and establish a JTF.[17] JTF 536 was formally established the next day, December 28, 2004, to conduct Operation Unified Assistance. To emphasize the collaborative nature of the operation, which involved forces from 21 countries, the JTF was subsequently designated a "Combined Support Force" 536 (CSF).

Many of the countries involved had already built relationships with one another through their work in the Multinational Planning Augmentation Team (MPAT), which was created in November 2000 by the PACOM commander and defense chiefs of several Pacific-Asian

---

[15] Tsunami Evaluation Coalition, *Joint Evaluation of the International Response to the Indian Ocean Tsunami: Synthesis Report*, London, U.K.: Active Learning Network for Accountability and Performance in Humanitarian Action, July 2006, p. 37.

[16] "PACOM PLANORD [Planning Order], 271009Z Dec 04, HQ PACOM to COMARFORPAC [commander of the Army Forces in the Pacific]," quoted in "Operation Unified Assistance Chronology."

[17] "PACOM PLANORD [Planning Order], 271115Z Dec 04, HQ PACOM to COMARFORPAC [commander of the Army Forces in the Pacific]," quoted in "Operation Unified Assistance Chronology."

countries. MPAT was created to establish standard operating procedures for operations for low-intensity contingencies and to maintain a group of planners who could augment a "combined task force headquarters" during a crisis. In addition to personnel from partner nations, MPAT includes representatives from international and nongovernmental organizations.[18] U.S. military personnel reported that the operation benefited from exercises, such as Cobra Gold, with foreign military partners.[19] Senior DoD leaders agreed.[20]

In preparation for a humanitarian crisis, PACOM had designated I Corps, Seventh Fleet, Third Fleet, I MEF, and III MEF as potential JTF headquarters under Contingency Plan 5070-02, *Foreign Humanitarian Assistance (FHA) and Peacekeeping (PK)/Peace Enforcement (PE) Operations*.[21] After the tsunami, III MEF was tasked with forming the JTF headquarters. It was led by Lt Gen Robert R. Blackman, Jr. The command element was deployed forward to Utapao, Thailand. JFACC was located at Hickam Air Force Base in Honolulu, Hawaii. The U.S. portion of the CSF headquarters consisted of 986 personnel, 66 percent of whom were Marines. See Figure 4.2 for more detail.

---

[18] Multinational Planning Augmentation Team, "Multinational Planning Augmentation Team (MPAT) and Multinational Force Standing Operating Procedures (MNF SOP) Programs," Information Paper, PACOM J722, October 1, 2007b; Multinational Planning Augmentation Team, "Multinational Planning Augmentation Team (MPAT): What Is MPAT?" briefing, October 1, 2007a; and Multinational Planning Augmentation Team, "Multinational Force Standing Operating Procedures: Overview Brief," January 1, 2008. An officer from Singapore reported that MPAT provided a valuable opportunity to integrate U.S. and other militaries. See Col Mark Koh, Singapore Armed Forces, "Operation Unified Assistance—A Singapore Liaison Officer's Perspective," *Liaison*, Vol. 3, No. 3, 2005.

[19] "U.S. Military Relief Efforts for Tsunami Victims," CAPT Rodger Welch, USN, Tsunami Relief Spokesperson, U.S. Pacific Command, Camp H. M. Smith, Hawaii, January 5, 2005.

[20] See ADM Thomas Fargo and Deputy Secretary of Defense Paul Wolfowitz, both quoted in Ralph A. Cossa, "South Asian Tsunami: U.S. Military Provides 'Logistical Backbone' for Relief Operation," *e-Journal USA, Foreign Policy Agenda*, November 2004, updated March 2005.

[21] Lt Gen David A. Deptula, *CSF-536 Joint Force Air Component Commander (JFACC)/Air Force Forces Commander (AFFOR) Lessons and Observations*, Pacific Air Forces, May 17, 2005, p. 17.

Figure 4.2
Composition of Combined Support Force–536 HQ, 986 Total U.S. Personnel

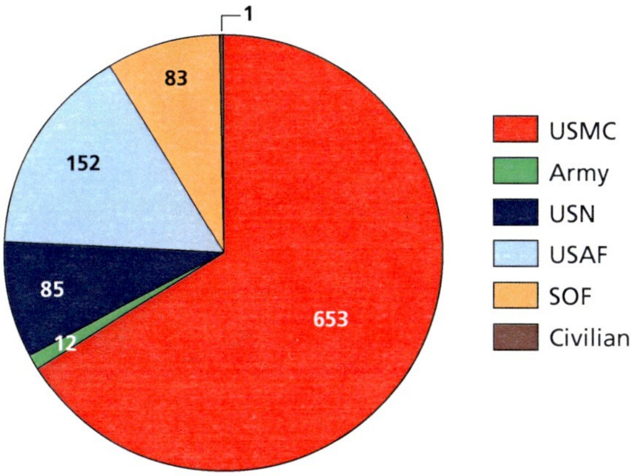

- USMC: 653
- Army: 12
- USN: 85
- USAF: 152
- SOF: 83
- Civilian: 1

RAND MG777-4.2

Lack of warning made planning difficult. The guidance that PACOM sent out immediately following the tsunami set portions of the military response in motion, most notably the Disaster Response Assessment Teams (DRATs), which were each led by a colonel and deployed to Sri Lanka, Indonesia, and Thailand to gauge the scope of the humanitarian response required. The teams consisted mostly of Marines. While these teams began their work, getting other assets in place lagged because of vague initial guidance. For example, instructions for the Air Force to deploy between five and eight C-130s failed to specify any kind of executable deployment strategy. Without guidance on where the hub of operations would be (the execute order was issued prior to the decision to base out of Utapao, Thailand), it was difficult to mobilize the aircraft, and as late as the evening of December 28, 2004, those aircraft could not fly into Utapao because there was no plan of employment for them once they arrived.[22]

---

[22] Headquarters, Pacific Air Forces (PACAF), *With Compassion and Hope: The Story of Operation Unified Assistance: Air Force Support for Tsunami Relief Operations in Southeast Asia 25 December 2004–15 February 2005*, Hickam Air Force Base, Hawaii: PACAF, Air Education and Training Command's Historian office, not dated, pp. 15–16.

Lieutenant General Blackman set forth his operational goals in a base order issued January 5, 2005. He divided the operation into four phases: deployment and assessment, execution, transition to IOs, and redeployment.[23] The first phase was well under way by this point, positioning assets in-theater and establishing a headquarters at Utapao. Phase I specified that JFACC would coordinate with DRATs for tasking.[24] Phase II transitioned DRATs into Combined Support Groups (CSGs) with tactical control of their AORs and tasked the JFACC with inter- and intratheater airlift. There were three CSGs designated to Indonesia (CSG-IN), Sri Lanka (CSG-SL), and Thailand (CSG-TH). Each CSG was eventually led by a Marine brigadier general. This meant that CSF-536 consisted of medium-based air, land, and sea components; service components that provided forces; and geographical components devoted to particular countries. The primary command elements and the relationships among them are shown in Figure 4.3.

This left a somewhat vague relationship between DRAT/CSG structures and JFACC. This was only somewhat clarified in the base order, in which Lieutenant General Blackman noted:

> CSG-TH, CSG-SL and CSG-IN are the tactical supported commanders. JFACC is the supported commander for air mobility, ISR, and tactical lift missions within the affected areas of Sri Lanka, Thailand, Indonesia, and India (if necessary); with the exceptions for NAVFOR, JFSOC [Joint Force Special Operations Command], and CSG's noted below. MARFOR [Marine Corps Forces], AFFOR, NAVFOR, and JFSOC are the tactical supporting commanders. . . . NAVFOR retains tasking authority of organic aircraft. All other aircraft are assigned TACON [tactical control] to JFACC for tasking by JTF-536.[25]

---

[23] Joint Task Force 536, *Base Order,* Washington, D.C.: The Joint Staff, January 7, 2005.

[24] JTF 536, 2005.

[25] JTF 536, 2005.

Figure 4.3
Combined Support Force–536 Organizational Chart

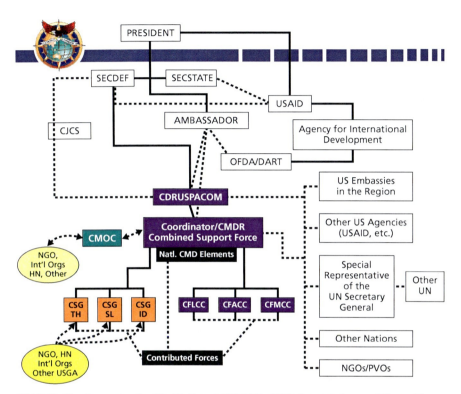

SOURCE: Headquarters, Pacific Air Forces (PACAF), *With Compassion and Hope: The Story of Operation Unified Assistance—Air Force Support for Tsunami Relief Operations in Southeast Asia 25 December 2004–15 February 2005*, Hickam Air Force Base, Hawaii: PACAF, Air Education and Training Command's Historian office, not dated.
RAND MG777-4.3

This confusing command and control relationship was due in part to the lack of warning of the operation, but also to the need for the operation to address the varying needs of the Indonesian, Sri Lankan, and Thai people in three geographically distinct locales.

CSGs functioned as small geographic commands with control over all assets in their AOR. JFACC, on the other hand, wanted tactical control of all air assets, fixed and rotary, regardless of location. To complicate matters further, the naval component sought to retain tactical control of its organic aircraft. The command and control relationships were not clarified until Blackman issued Fragmentary Order (FRAGO) 6 on January 15. That order stated the following:

> 12.b. AFFOR, MARFOR, NAVFOR, and JFSOC are the tactical supporting commanders and retain OPCON [operational control] of assigned forces. TACON is inherent in OPCON unless delegated in para[graph] 12.d.
>
> 12.c. CSG-I, CSG-T and CSG-SL are the tactical supported commanders.
>
> 12.d. . . . CFACC is assigned TACON of all CSF-536 U.S. aircraft for tasking with the following exceptions:
>
> . . .
>
> 12.d.4. All other land-based R/W [rotary wing] squadrons will be assigned direct support to a CSG, and remain TACON to their component Commander.
>
> 12.d.5. Other sea-based R/W squadrons remain TACON to NAVFOR/MARFOR. NAVFOR/MARFOR retain tasking authority of aircraft and, as the supporting commanders, respond to the support requirements of the supported CSG's.[26]

The rear location of the JFACC contributed to another command-and-control shortcoming. Though an Air Component Coordination Element was forward deployed, the JFACC remained at Hickam to test its warfighting headquarters concept in the newly expanded AOC. This created a time lag in requests for scheduling, which confounded some operations.[27] Similar to operations in the CENTCOM AOR, there is a question of whether it is better for the Air Force to provide one AOC for a theater in which there are several operations ongoing simultaneously, or if another command arrangement would be more

---

[26] JTF 536, 2005.

[27] Headquarters, Pacific Air Forces, not dated, p. 39.

appropriate.[28] Regardless of whether a Marine Corps or an Air Force unit leads the operation, this question would remain.

How might CSF-536 have differed if it had been led by the Air Force? It is unlikely that an Air Force–led headquarters would have mixed medium-based components with geographical and service components. The tenet of centralized control and decentralized execution indicates a preference for one command entity that can allocate forces to different areas across the area of operations. Without having command elements dispersed, it might have been more difficult to address different conditions in Indonesia, Sri Lanka, and Thailand. On the other hand, a more centralized approach might have reduced confusion and led to more-efficient and more-effective operations.

Since the U.S. military's ability to deploy lift and assessment capabilities quickly and in large quantities is unparalleled, CSF-536's involvement in the tsunami response was welcomed by the international community. There were some concerns about a lack of integrated planning between the military and nonmilitary actors, as well as criticism that U.S. forces lacked knowledge of local political conditions.[29] Nevertheless, it is clear that U.S. forces helped to ease human suffering and to promote a positive American image.

Like Operation Atlas Response, CSF-536 shows the importance of coordinating with local partners. The nondoctrinal use of a CSF instead of a JTF shows the importance that decisionmakers placed on having the U.S. military show its willingness to work alongside, rather than over, others. We also see the importance of advance planning efforts and establishing relationships among partners prior to the operation.

---

[28] Bob Poynor, "Is Air Force Command and Control Overly Centralized?" Montgomery, Ala.: Maxwell-Gunter Air Force Base, Air University, June 20, 2007.

[29] Tsunami Evaluation Coalition, 2006, pp. 99, 60.

JTF–Noble Anvil: The Questionable Joint Task Force

OAF is an interesting event for the Air Force to consider. The operation to counter Serbian aggression against Kosovar Albanians relied mostly upon U.S. Air Force aircraft, yet the JTF commander for the operation was a Navy admiral working for a combatant commander who was an Army general. Another curious feature was that the operation included two parallel chains of command: one U.S.-only and one North Atlantic Treaty Organization (NATO). OAF is remembered as an example of how difficult allied operations can be, because military decisionmaking was significantly constrained by alliance considerations. To maintain the cohesion of the alliance and popular support for the intervention, political leaders forswore the use of ground troops and even forbade significant efforts to draw up plans that would feature their participation. Allied governments also inserted themselves into the details of the targeting process, adding to the frustration of military commanders.[30]

JTF–Noble Anvil (JTF-NA) was established in January 1999 under the command of ADM James O. Ellis, who at the time was Commander-in-Chief, U.S. Naval Forces, Europe, and Commander-in-Chief, Allied Forces, Southern Europe. JTF-NA was the U.S. component to OAF, which was also commanded by Admiral Ellis. Military strikes against Kosovo began on March 24, 1999, and ended when Serbia withdrew its forces from Kosovo 78 days later. The core of Admiral Ellis' staff was made up of U.S. Naval Forces, Europe, personnel stationed in London. The forces deployed to Naples, Italy, for the operation. This was an administrative staff that had not been trained to

---

[30] For more detail, see Bruce R. Nardulli, Walter L. Perry, Bruce R. Pirnie, John Gordon IV, and John G. McGinn, *Disjointed War: Military Operations in Kosovo, 1999*, Santa Monica, Calif.: RAND Corporation, MR-1406-A, 2002; Benjamin S. Lambeth, *NATO's Air War for Kosovo: A Strategic and Operational Assessment*, Santa Monica, Calif.: RAND Corporation, MR-1365-AF, 2001; and Ivo H. Daalder and Michael E. O'Hanlon, *Winning Ugly: NATO's War to Save Kosovo*, Washington, D.C.: Brookings Institution Press, 2000.

plan and oversee military operations.³¹ The JMD for the Noble Anvil headquarters lists 326 positions, as shown in Figure 4.4. Of these 113, or 35 percent, were filled by naval personnel. There were relatively large contingents of Air Force (25 percent) and Army (20 percent) personnel, as well. Fortunately for the JTF headquarters staff, a good deal of planning had previously been undertaken by two other JTF headquarters. Planning began in May 1998 for a possible contingency against Serbia.³² U.S. planners came up with a plan for a series of air strikes, called Nimble Lion. NATO planners also drew up a plan, Concept of Operations Plan (CONOPLAN) 1061, that envisioned a more incremental approach.³³ From August to December 1998, DoD established two JTFs to help prepare for possible contingencies against Serbia. JTF–Flexible Anvil, under the command of VADM Daniel J. Murphy and the U.S. Navy's Sixth Fleet, was set up to plan limited strikes using conventional air-launched cruise missiles and Tomahawk land-attack missiles (TLAMs). JTF–Sky Anvil, commanded by Lt Gen Michael Short and the Air Force's Sixteenth Air Force, contemplated the use of fixed-wing aircraft in the event that JTF–Flexible Anvil's strikes were insufficient.³⁴ Flexible Anvil and Sky Anvil were both disbanded in December 1998 when Ambassador Richard Holbrooke persuaded Serbian President Slobodan Milosevic to allow observers from the Organization for Security and Co-operation in Europe into Kosovo and NATO flights to verify troop movements.³⁵ CONOPLAN 1061 would eventually form the basis for OAF, which was initiated after the situation in Kosovo worsened.

---

[31] ADM James O. Ellis, USN, "A View from the Top," briefing, October 21, 1999; and CAPT J. Stephen Hoefel, USN, *U.S. Joint Task Forces in the Kosovo Conflict*, Newport, R.I.: U.S. Naval War College, May 16, 2000.

[32] Lt Gen (ret) Michael Short, quoted in U.S. Senate, *Hearing of the Senate Committee on Armed Services: Lessons Learned from Military Operations and Relief Efforts in Kosovo,* October 21, 1999, p. 8.

[33] Lambeth, 2001, p. 11.

[34] U.S. Department of Defense, *Report to Congress: Kosovo/Operation Allied Force After-Action Report*, January 31, 2000, p. 18.

[35] Hoefel, 2000, p. 5.

**Figure 4.4
Composition of JTF–Noble Anvil HQ, 326 Total Personnel**

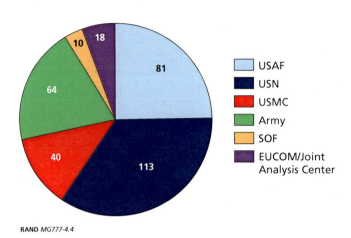

RAND MG777-4.4

While NATO's North Atlantic Council issued three "strategic objectives" for the conflict, our research has been unable to locate any operational-level goals that could have been used to organize allied efforts.[36] The closest available substitute was the manner in which the campaign was divided into phases. Phase 1 focused on the need to take down Serbian air defenses and to establish air superiority over Kosovo. Phase 2 of the operation, which commenced on the fifth day of hostilities, dealt with strikes on Serbian military forces in Kosovo and south of the 44 degrees north latitude. It emphasized interdiction of Former Republic of Yugoslavia (FRY) forces and lines of communication rather than suppression of air defenses.[37] Phase 3, which began nine days into the conflict, expanded attacks on FRY leadership, command and control nodes, and other targets, including some in Belgrade.[38]

---

[36] The three strategic objectives were to (1) "demonstrate the seriousness of NATO's opposition to Belgrade's aggression in the Balkans," (2) "deter Milosevic from continuing and escalating his attacks on helpless civilians and create conditions to reverse his ethnic cleansing," and (3) "damage Serbia's capacity to wage war against Kosovo in the future or spread the war to neighbors by diminishing its ability to conduct military operations." All quotes are from U.S. Department of Defense, 2000, p. 7.

[37] Lambeth, 2001, p. 25.

[38] Lambeth, 2001, p. 29.

The problem with the planning for Allied Force was not that there was not enough planning, but rather that planning was highly constrained. Clinton administration officials explicitly opposed talk of sending in ground forces. National Security Advisor Samuel Richard "Sandy" Berger argued that refraining from using ground troops was necessary to maintain allied support for the operation.[39] Berger was also uncertain whether the U.S. Congress would support the use of ground forces.[40]

Planning was also constrained by a poor appreciation of Serb motivations. Many remembered that Milosevic and his Bosnian Serb allies capitulated after less than a month of bombing during Operation Deliberate Force in 1995, and they therefore expected a short campaign. Even after the campaign began, NATO Secretary General Javier Solana expressed certainty that the conflict would end before NATO's 50th anniversary in late April.[41] After the conflict, GEN Wesley Clark explained:

> It wasn't planned by the nations as a war, an all-out war against Milosevic. They couldn't see it that way. They saw it as, Maybe we can just show that we're very serious and he'll come back to the bargaining table.[42]

Admiral Ellis, the JTF commander, noted that "[w]e called this one absolutely wrong," noting that the operation lacked coherent campaign planning, lacked adequate component staffing, and failed to consider alternative outcomes and different courses of action.[43]

To the extent that JTF headquarters are intended to orchestrate the various force elements available to them, one may rightly question whether JTF-NA was really a "joint" task force. There were some TLAM strikes from the sea, Navy aircraft electronically suppressed Ser-

---

[39] Daalder and O'Hanlon, 2000, p. 97.

[40] Nardulli et al., 2002, p. 23.

[41] Lambeth, 2001, p. 43.

[42] GEN Wesley Clark, quoted in U.S. Senate, 1999, p. 12.

[43] Ellis, 1999.

bian radar and flew strike missions, and special operators did provide combat search and rescue capabilities. General Clark worked diligently to incorporate Apache helicopters into operations but was delayed by the Army's concerns that they would be used in a way that ran counter to Army doctrine and pilot training.[44] The Army's Hunter unmanned aerial vehicles and counterfire radars were employed to help AOC find targets.[45] Nevertheless, the predominance of strikes were delivered by land-based, fixed-wing aircraft. The *Theodore Roosevelt*, a carrier, was sent out of the area to travel to the Persian Gulf prior to the conflict, and it did not return until two weeks into the fight.[46]

OAF raises the question of why JTF-NA, a U.S.-only command, was created in addition to having a NATO chain of command. The United States maintained the separate ATO for assets such as F-117s, B-2s, and TLAMs, and it has been maintained by some that JTF-NA was created to provide command and control of these assets, to protect classified material, and to provide a link between classified assets and U.S. allies.[47] Others charge that these functions could have been performed at the Combined Air Operations Center in Vicenza and that the JTF was formed as a way for General Clark to directly manage the use of B-2s and other selected assets.[48] Vehement disagreements between Generals Clark and Short about the best way to employ air power seem to support this argument. General Clark thought that priority should be given to striking FRY forces in the field, while Lieutenant General Short argued that it would be better to go "for the head of

---

[44] For more information on Task Force Hawk, the attempt to incorporate the Apaches, see Nardulli et al., 2002, Chapter Four.

[45] Nardulli et al., 2002, p. 90.

[46] VADM Daniel J. Murphy, USN, "The Navy in the Balkans," *Air Force Magazine*, Vol. 82, No. 12, December 1999.

[47] Lt Col Paul C. Strickland, USAF, "USAF Aerospace-Power: Decisive or Coercive," *Aerospace Power Journal*, Fall 2000; and Maj. Antonio J. Morabito III, USMC, *NATO Command and Control: Bridging the Gap*, Newport, R.I.: Naval War College, February 5, 2001, p. 6.

[48] Interview with Lt Gen (ret) Michael Short, July 17, 2007.

the snake on the first night," or, in other words, to pursue targets such as the power grid and leadership nodes in Belgrade.[49]

JTF-NA serves as a vivid reminder of just how difficult it can be to run a major military operation. Keeping alliances and coalitions together often requires a sacrifice in military effectiveness. Politicians can and will affect military decisionmaking. Combatant commanders often disagree with their subordinates about how to run an operation. These are the sort of challenges that Air Force JTF commanders will face in the future.

JTF–Southwest Asia: *Groundhog Day*[50]

Operation Southern Watch began in the wake of Operation Desert Storm. The United Nations Security Council (UNSC) passed Resolution 688 in April 1991 demanding an end to Saddam Hussein's brutal repression of Iraqi civilians. In 1992, the UN determined that Hussein had not complied with the terms of UNSC Resolution (UNSCR) 688. In support of this finding, on August 19, 1992, the Chairman of the Joint Chiefs of Staff issued an alert and deployment order. One week later, President George H. W. Bush announced the creation of the Southern Watch NFZ and the establishment of Joint Task Force–Southwest Asia (JTF-SWA) under Lt Gen Michael A. Nelson. The next day, JTF-SWA flew its first sortie in support of the new mission. In all, there were only eight days between warning and deployment, one of the shortest response times to establish one of the longest-running JTFs.[51] JTF-SWA remained operational until May 1, 2003.

---

[49] Lt Gen (ret) Michael Short, quoted in U.S. Senate, 1999, p. 8. Also see Strickland, 2000 and Dana Priest, "The Battle Inside Headquarters; Tension Grew with the Divide over Strategy," *Washington Post*, September 21, 1999, p. A1.

[50] Many of the personnel involved in Operation Southern Watch compared it to the movie *Groundhog Day* because of its routine nature. For example, see A. J. Plunkett, "U.S. Still Polices No-Fly Zone Over Southern Iraq: Monotony Rules 4-Year Mission," *Newport News Daily Press*, December 25, 1994; and "Battle Alert in the Gulf," transcript of *Nova*, Public Broadcasting System, aired February 23, 1999.

[51] Estrada, 2005, pp. 26, 31.

JTF-SWA fell under the overall responsibility of the U.S. Central Command and was staffed by U.S. Air Forces Central Command and U.S. Naval Forces Central Command (NAVCENT) personnel. It had a threefold mission:

- To plan and, if directed, conduct an air campaign against Iraqi targets as a means of compelling Iraq to comply with UNSCR 687, which calls for UN inspections of Iraqi weapons-making potential.
- To enforce the NFZ south of 33 degrees north in Iraq, in support of UNSCR 688, demanding Iraqi leader Saddam Hussein end his suppression of the Iraqi civilian population.
- To enforce a no-drive zone south of 32 degrees north in Iraq in support of UNSCR 949 to prevent enhancement of Iraqi military capabilities in southern Iraq.[52]

In addition to preventing the flight of any kind of Iraqi aircraft in the NFZ without prior UN approval, the JTF also sought to employ surveillance assets to ensure that the Iraqi Army was not making any significant movements. The JTF commander sought to do all of this with as little risk to JTF personnel as possible.[53]

Operation Provide Comfort, which began in July 1991, established another NFZ in northern Iraq. The operation transformed into Operation Northern Watch in 1997. Though Northern Watch and Southern Watch shared the common objective of restricting Iraqi air mobility, the operations differed. Southern Watch operated roughly four times the number of aircraft as Northern Watch, and it also covered a larger area.[54] Further, because of Unified Command Plan–defined areas of responsibility, Northern Watch was conducted under the auspices of EUCOM. As a CENTCOM effort of a large scale, Southern Watch represented new territory for CENTCOM and for Ninth Air Force.

---

[52] James Kitfield, "The Long Deployment," *Air Force Magazine*, Vol. 83, No. 7, July 2000.

[53] Interview with General Nelson.

[54] John T. Correll, "Northern Watch," *Air Force Magazine*, Vol. 83, No. 2, July 2000.

Figure 4.5 shows that the JTF headquarters was staffed primarily by the Air Force, with a deputy commander provided by NAVCENT. Though primarily an Air Force and Navy effort, JTF-SWA also utilized U.S. Army, Royal Air Force, and French air units.[55] For a time, participants served 90-day tours.[56] The size of the headquarters staff fluctuated between approximately 100 and 300. In July 1997, 74 percent of the 251-person staff was U.S. Air Force.[57]

The structure of the JTF headquarters strongly resembled that of JFACC organization from Desert Storm. The JTF-SWA HQ had a J-1 through J-6 structure, minus a J-5, which changed little throughout the course of the operation. In fact, JTF-SWA followed other

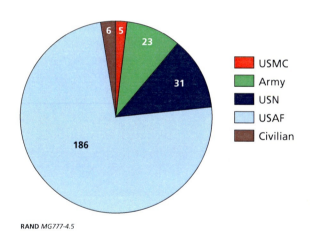

**Figure 4.5
Composition of JTF–Southwest Asia HQ, 251 Total Personnel**

RAND *MG777-4.5*

---

[55] Michael A. Nelson and Douglas J. Katz, "Unity of Control: Joint Air Operations in the Gulf—Part Two," *Joint Forces Quarterly,* Summer 1994, p. 60.

[56] In 1995, the commander's tour was extended to one year. Nelson and Katz, 1994, p. 61.

[57] Page 41 of the July–August 1997 unit history of the Ninth Air Force, Shaw Air Force Base, Sumter, S.C.

procedures established during Desert Storm, such as a single ATO for each flying operation, a single airspace control order and responsibility for area air defense.[58]

Operation Desert Storm also aided the planning process for JTF-SWA. A number of assets remained in theater, and many of the participants in the JTF headquarters had worked together in Desert Storm.[59] An example of this was the planning for the use of Army Patriot batteries, one of the few non-air assets used by the JTF. There was a considerable amount of coordination between the Patriot operations and the ATO, as well as personally between the JTF commander and the colonel who commanded the Patriot forces.[60] That this coordination proceeded smoothly was largely the result of professional relationships and prior experience dealing with these assets during Desert Storm. Another factor that eased planning and increased readiness was the fact that, despite the short official warning time to establish the JTF, General Nelson and his staff were already in theater, having been asked to begin discussions with the Saudis in anticipation of a U.S.-led response. Because of the presence of assets in the region, the prior history of service members there, and the unofficial advance team for the JTF, planning for Southern Watch proceeded very smoothly.

The JTF-SWA headquarters functioned well as a joint organization. With nearly a decade to tinker with its model, it was able to successfully integrate the primarily Navy augmentees into its structure. In addition to the staff assigned to the headquarters on a standing rotational basis, there were also naval liaison officers from battle groups and Navy air wings assigned to support Operation Southern Watch.[61]

While JTF-SWA staff functions were joint, the predominant role of aircraft precludes it from serving as a model for future Air Force–

---

[58] Nelson and Katz, 1994, p. 60. The airspace control order, part of the guidance issued by the air component commander, sets procedures to prevent aircraft from colliding with one another.

[59] Author's interview with Lt Gen Michael A. Nelson, August 22, 2007.

[60] Author interview with General Nelson.

[61] LCDR Nicholas Mongillo, USN, *Navy Integration into the Air Force–Dominated JFACC*, Newport, R.I.: Naval War College, February 8, 2003, p. 5.

led JTF headquarters. JTF-SWA was essentially a large, well-directed CFACC, without a Combined Force Land Component Command (CFLCC) or CFMCC. There were Patriot batteries manned by the Army, but in general, the JTF commander's role was to orchestrate air operations, rather than to harmonize the full array of air, land, and sea operations. Another interesting aspect of Southern Watch is that the mission was to maintain a steady-state of calm.[62] Thus, the daily pattern of activities remained relatively static. While this contributed to a smoothly operating headquarters, it does not replicate the chaotic nature of combat operations that will test future JTF headquarters.

Summary

These JTFs are typical in several ways. In the two examples of humanitarian operations, there was little or no warning. All of the JTFs discussed above show the need to build and maintain good relationships with partner militaries and other non-DoD organizations. Operation Southern Watch shows how JTF operations can extend for long periods of time.

The operations discussed above also indicate that despite being labeled as "joint" task forces, operations tended to be dominated by forces from one service and/or medium. JTF-AR, JTF-NA, and JTF-SWA were all dominated by the use of land-based aircraft, even though JTF-NA was nominally led by a Navy admiral (and overseen by an Army general). In none of these JTFs do we see much evidence of a commander who integrates elements from different services and mediums into much more than the sum of their parts. The "jointness" envisaged by so many is not in evidence here.

At the same time, it is important to point out that these operations did not require more jointness to be successful. Humanitarian operations are difficult to evaluate in terms of effectiveness—a JTF could be more or less efficient and effective at providing relief to people in distress, but the difference between success and failure is more a

---

[62] Interview with General Nelson.

matter of judgment and degree. In combat operations, assessment can be difficult, but there is usually some discernable evidence of success or failure. In OAF, there was Milosevic's capitulation, and in Operation Southern Watch, there was the lack of significant Iraqi efforts to challenge militarily the no-fly and no-drive zones.

The important issue for our purposes here is that if, in the future, U.S. military forces have to face a more capable adversary than the Former Republic of Yugoslavia or Saddam Hussein's Iraq, it is not clear that DoD would be able to produce a command entity that could orchestrate the operations of different types of forces into a coherent campaign. If jointness were necessary for success, would U.S. forces be able to respond to the challenge? These operations and other recent examples are not auspicious.

When asked about Air Force leadership of JTFs, air power advocates often point to the lack of jointness in JTF operations led by other services.[63] Regardless of the problems faced by others in integrating a joint force, the Air Force would do better to ask itself how it would do a better job.

---

[63] Multiple interviews with Air Force officers, 2006–2007.

CHAPTER FIVE
Requirements

Having surveyed relevant developments in the other services and elsewhere in the defense community, as well as examining a few examples of actual JTF headquarters, we are in a position to postulate a set of requirements that the Air Force would need to meet in order to lay the foundation for one or more JTF headquarters. There are three basic requirements: build, prepare, and execute.

Build

"Build" requirements are institutional-level efforts. There are four basic elements to a JTF headquarters core: a commander, a staff, a facility, and equipment. The latter two are necessary for a headquarters to be successful. However, they are relatively simple requirements to identify and fulfill. In fact, the Air Force has proven itself to be adept at both through its efforts to upgrade the capabilities of its AOCs. Thus, we will focus on the former two. To supply commanders and staffs, the military services must develop a pool of qualified officers. This is done through recruiting and selection, through training and education, and through career paths that expose candidates to experiences that will best prepare them for their posts.

Develop Commanders

There are many exceptional individuals in the higher ranks of the American military, but it is more prudent to devise a program of selection, instruction, and experiences than to rely on chance. The small number of leadership positions in relation to the size of the officer corps makes it difficult, if not impractical, to engineer the officer selection process to produce good candidates for JTF leadership. Individual training and education, however, both offer opportunities to provide individual officers with knowledge of the operational art; a capacity for absorbing, organizing, and prioritizing information; and an appreciation for the need to work with others.

Management of career paths offers another opportunity to create good JTF commanders. Ideally, a JTF commander will have had a chance to lead groups of people in the accomplishment of complex tasks. This requires the ability to define discrete goals from broad guidance, to set priorities, and to orchestrate available resources in the best possible manner.

Breadth of education and experience helps officers function in a multifaceted environment.[1] Experience working with other services, particularly with those who operate in different mediums, is invaluable for a commander who must employ different types of forces in combat. Moreover, it would greatly benefit a JTF commander to have had the experience of working outside the military with interagency and international partners. Unfortunately, these types of broadening experiences tend to take officers away from narrower service-related experiences that often offer better prospects for promotion.

Build Staffs
Similar to building JTF leaders, building JTF headquarters staff requires the services to select, train, educate, and manage the careers of commissioned and noncommissioned officers. Again, the small proportion of personnel who serve in JTF headquarters in relation to the general force is such that it is not practical to gear recruitment and selection beyond procuring the best candidates among those available. With respect to education and training, staff candidates should gain expertise in some area of military operations, an appreciation for operational art, and an

---

[1] Col Mike Worden, USAF, *Rise of the Fighter Generals: The Problem of Air Force Leadership, 1945–1982,* Maxwell Air Force Base, Ala.: Air University Press, March 1998. This is one of the major themes of (now) General Worden's book.

appreciation of the need to work with interagency and international partners. It is also important for staffs to receive training specific to the types of tasks they would be asked to perform on a staff. For example, staff who work with personnel issues would ideally receive training on how to assess the need for, how to request, and how to obtain personnel from other services through the JMD process.

A formal training program is necessary to produce competent staffs. Such a program, geared toward teaching individuals how to carry out various staff functions, generates a pool of capable commissioned and noncommissioned officers. As part of the program, a certification process would ensure that personnel meet set standards of performance. Continuity training would help refresh the skills of individuals who have not had recent experience on a relevant headquarters staff.

With respect to career paths, personnel who have had prior experience working on a headquarters staff, particularly a JTF headquarters, would be ideal. Like commanders, staff members who have had the opportunity to broaden their perspectives through direct association with likely joint, interagency, and international partners would help the JTF staff work more effectively. Having the ability to track people with these types of experiences would help the Air Force locate them quickly if they are needed. Lastly, promoting people with this type of experience is necessary to maintain a cadre of motivated, competent personnel.

Prepare

While the institutional Air Force is responsible for building JTF headquarters capability, it is the prospective JTF headquarters staff that must take action to prepare for the role. Preparation requires persistence, foresight, and engagement.

Identify Missions

JTF headquarters staffs need to prepare by identifying their likely missions. Commanders and staffs are usually given missions and do not get to choose them. Nevertheless, Operation Atlas Response shows that identifying in advance a likely type of mission, preparing for it, and letting superiors know about a unit's readiness make it more likely for a unit to be named a JTF headquarters.

Identifying a type of mission in advance also helps focus planning efforts. Drafting and revising plans for the most likely and most critical potential operations are no small undertakings. Systematic efforts that consider a wide range of branches and sequels pay enormous dividends by saving staff time and effort during an operation. Planning also helps commanders consider the feasibility of various courses of action and identify necessary capabilities in advance of actual employment.

Exercise

To prepare, a commander and staff should also engage in appropriate exercises. These exercises should be as realistic as possible, based on the most likely or most critical missions that the headquarters would be given. They should involve likely interagency and international partners. In the case of HUMRO, JTF–Atlas Response and CSF-536 show the benefit of prior contact with NGOs and partner militaries. The exercises should seek to establish familiarity with the tasks at hand and with the steps necessary to achieve them. They can be used to increase staff proficiency and to identify gaps in staff capabilities so they can be addressed. Exercises should also be used to build relationships among the staff members and between the staff and likely partners as much as possible prior to an actual operation.

Engage Partners

Another required task that a would-be JTF headquarters staff can work on in advance of an operation is to engage likely partners. As in the identification of missions, JTF commanders and staffs do not generally pick their partners. Nevertheless, identifying likely partners and forging relationships with them through exercises or other interactions prior to an operation can help ease communications between partners, provide additional information about threats and opportunities, and build the capacity of partners to provide for their own security. Much of this work is done by combatant commands, often through service components. OAF shows the importance, as well as the difficulty, of engaging partners. Despite the constraints that foreign partners placed on the operation, success would have been extremely unlikely if the coalition had dissolved. OAF was an anomaly in that it was executed by a preexisting alliance. PACOM's MPAT also shows how advance planning with others can help build relationships with likely partners.

Execute

Similar to the requirements involved in preparing the JTF headquarters, the last set of requirements also falls upon the JTF headquarters itself and not the service. To execute involves the actual use of the JTF headquarters in an operation.

Build and Maintain Partnerships

Whether or not an operation involves working with a coalition, there are likely to be actors other than the U.S. Armed Forces playing important roles. Whether they are from other U.S. government agencies, partner militaries, partner governments, or NGOs, establishing and sustaining productive relationships between a the JTF and others often consume JTF commanders and their staffs. There are some who express the view that this sort of outreach activity is an onerous task that constitutes a distraction from the military's "real work," but this remains a critical task for a JTF headquarters. During OAF, Lieutenant General Short, the joint force air component commander, understandably expressed frustration about the limitations placed upon him by the need to maintain alliance unity, but as JTF commander, he would still have had to maintain allied solidarity. For Air Force officers who have traditionally focused on the role of the air component, stepping up to the JTF-headquarters level will necessitate the mastering of new skills such as this.

Staff the Headquarters

Throughout this monograph, we have referred to units that form the "core" of the JTF headquarters. To fulfill the functions of a JTF headquarters, core staff must be augmented by personnel from the unit's service and personnel from other services. Previous experience with JTF headquarters indicates that filling required positions with qualified personnel is a difficult task, and one that often takes as much as six months to complete.[2] Frequently, units can turn to SJFHQ (CE), but that will mean the addition of only 57 people, and the JMEP does not provide for large numbers of personnel, leaving the JMD process as the only other alternative. Staff members who know how to work the JMD process effectively can mean the difference between a fully functional staff and a skeleton headquarters.

Issue Orders

To accomplish the mission of the JTF, the JTF headquarters has to issue orders. This might seem obvious, but the process of issuing formal orders is, in some places in the U.S. military, something of a lost art. There are reports that some NAF staffs do not have any personnel who know how to write a formal order.[3] With the ubiquity of PowerPoint, email, and video conferencing, much tasking is done informally, bypassing what is seen to be a cumbersome process. While this may be the case, the process of drafting formal orders requires staffs to think through issues at a level of detail that is lost in more expeditious methods. Formal orders are also less likely to be misunderstood by personnel from other services and countries.

Gain and Maintain Situational Awareness

To function effectively, a JTF headquarters needs to be able to gather and process information about the environment. To do this, it needs to have access to the products of capable sensors and staffs that assess these products, as well as sufficient communications capability to carry information to and from sensors and headquarters. In addition to hardware, it needs analysts who are capable of interpreting data and discerning meaning from large amounts of information. The Air Force works with a wealth of airborne and space-based sensors, and it has experience moving and interpreting large amounts of data, so this

---

[2] Bonds, Hura, and Young (forthcoming); Estrada, 2005.

[3] Author interviews with former AFFOR Chief of Staff and other officers.

should be an area of relative strength for the service. Still, to function as a JTF headquarters, a NAF headquarters would need to incorporate sensor traffic and analysis from other services and government agencies into its operations.

Orchestrate Efforts

The JTF headquarters works at the operational level of war. As we discussed above, effective operational-level commanders bring together the means at their disposal to achieve their goals. One of the assumptions and aspirations of the drive to make the U.S. Armed Forces more "joint" is that the capability generated by different services and mediums working together will be greater than that from uncoordinated, individualized efforts. Several of the operations we consider above indicate that JTFs often rely mostly on forces from one service and/or medium, despite their designation as joint entities. For example, Operations Allied Force and Southern Watch were carried out mostly by land-based aircraft. Operation Atlas Response was also dominated by Air Force personnel and platforms. Other examples of JTFs also offer similar examples of dominance by one service or medium. As we see from JTF-NA, this is not just a feature of Air Force–led JTFs. Even today there are complaints that the U.S. headquarters in Iraq, Multi-National Force–Iraq (MNF-I) and Multi-National Corps–Iraq (MNC-I), are less like JTF headquarters and more like Army headquarters. It is important to note that in each JTF we examined earlier, the appropriate tools were used for the mission, and these operations were all considered to be successful ones. It is more important to accomplish the mission than to distort operations in the name of jointness. It is instructive, however, that more than 20 years after the passage of the Goldwater-Nichols Act, JTFs often lack "jointness." The real requirement is that when it is necessary to combine forces and formulate strategies for integrated operations, commanders and their staffs must be ready to do so effectively.

For Air Force officers to lead a JTF headquarters, they must be competent at incorporating the contributions of forces operating on the ground and at sea. An Air Force–led JTF headquarters would also need to incorporate commissioned and noncommissioned officers from other services into a number of roles within the headquarters to take advantage of their expertise and to ensure that strategy and operations reap the benefit of a diversity of perspectives.

Assess and Adjust

The last requirement deals with the JTF headquarters' ability to incorporate new information and to respond appropriately. This requirement is related to the need to gain and maintain awareness, but it deals more with how the JTF headquarters incorporates that information and acts on it. Here it is useful to recall Helmuth von Moltke the Elder's admonition that plans rarely survive first contact with the enemy. When conditions warrant, JTF commanders and staffs must determine that a change of course is necessary, adjust strategy, and then communicate those changes to fielded forces.

OAF showed the importance of being able to adjust to changes when, against expectations, Milosevic did not capitulate after the first few days. Because of the air-centric nature of the operation, the burden of adjustment fell more upon the air component than on the JTF headquarters, however. The CFACC commander and his staff had to develop a longer-term air campaign more or less on the fly.

This is not a purely reactive characteristic. For example, in Operation Atlas Response, General Wehrle and his staff established metrics based on the availability of basic infrastructure and the resettlement of displaced people early in the operation in order to determine when U.S. military efforts should end. This is also not a call to mindlessly pursue an inappropriate end state. Quite the opposite; this is about the ability to be able to respond to changing circumstances with agility.

CHAPTER SIX
Issues

There are a number of issues that the Air Force will need to consider as it builds the capability to form the cores of JTF headquarters. Some of these issues relate to any effort to create an operational-level staff. Others arise from unique aspects of the Air Force. We have chosen to separate these issues from the next chapter on recommendations. Some of the changes that the Air Force will need to make should emerge from discussion and debate within the institution.

Separate or Combine Employment and Management

To build JTF headquarters, the Air Force must carefully consider the division between force management and force employment. In the absence of an operation, most of the work available is by nature related to management. In the Air Force, ambitious and motivated personnel have gravitated toward this work to be able to showcase their productivity and value to their superiors and their peers. When an operation arises, these individuals will gravitate toward that as the arena in which to prove their worth. The problem with this model is that it reinforces some of the tendencies that the defense community has sought to correct in recent initiatives dealing with JTF headquarters, such as a lack of ready staffs, the need to assemble a team that has not yet "jelled," and a lack of knowledge and familiarity with the physical and cultural climate in which operations take place. Combining employment and management makes it easier to field staffs of high-quality officers, at a cost of lower readiness. Separating employment and management increases readiness, but risks having officers with high potential shun employment-focused duties in the absence of an operation.

As we have seen in our survey of the other services, each has made different choices with respect to the question of management versus employment. Army and Marine Corps staffs generally do not distinguish between those who work on employment and those who work on management tasks. The Navy focuses on the distinction between management and employment, but in practice, it has both combined and separated its staff work in these areas. The Air Force's C-NAF initiative indicates an institutional preference to separate management from employment as much as possible. First, as much management as possible is given to the "management MAJCOM." Then, within the C-NAF itself, the AFFOR staff works on more-immediate management issues while the AOC works on employment.

This initiative seems to indicate that the Air Force will continue to seek separation between employment and management as it considers JTF headquarters. The problem is that there is something of a mismatch between having the AFFOR staff focus on management in the absence of operations and then ask it to make a considerable shift to employment during a contingency. A management-focused staff is unlikely to be ready to orchestrate operations. Since AFFOR staffs focus primarily on Air Force issues and processes, adopting the JTF headquarters core role will require a considerable shift in duties and responsibilities. Adding a shift from management to employment will increase the difficulty.

Organize Around Time or Function

The further removed a staff is from fielded forces, the less it needs to be organized around the battle rhythm. JTF headquarters staffs are by no means immune to the importance of time, but they do not generally engage in minute-by-minute control over fielded forces. As we have seen, their focus is more on setting goals, orchestrating the operations of forces, maintaining relationships, and adjusting to change. The adoption of the J-code structure for AFFOR staffs indicates that the Air Force would likely adopt a similar structure for AFFOR-based JTF headquarters. It is interesting to note, however, that Air Force headquarters are moving toward the Napoleonic J-codes just as others, such as Army and Navy staffs, are moving away from this model.[1] It is possible that J-codes will work well for an Air Force JTF headquarters core. Nevertheless, the Air Force would do well to consider carefully the benefits and drawbacks of different functionally based organizational schemes before settling on an established construct.

Determine When the Air Force Leads a JTF and How Many Types of JTF Headquarters Does It Need

The Air Force is a good candidate for JTF leadership when operations rely heavily on land-based aircraft. Another indicator that an Air Force organization could form the JTF core would be when an operation takes place over large distances.[2] One question that looms large, of course, is whether it would be appropriate for the Air Force to lead an operation with a significant ground component or a major joint force combat operation. There are arguments that U.S. Air Forces have made significant advances in their ability to conduct major combat and that they are capable of shouldering more of the warfighting burden than they have in the past.[3] If this is true, then it would suggest that it is not out of the realm of possibility to consider Air Force JTF leadership in larger-scale operations. Of course, it is important to remember here that for operations of considerable size, such as Desert Storm and the first phase of Operation Iraqi Freedom, regional combatant command-

---

[1] One of the motivations for the Air Force to organize around J-codes was the desire to be able to interface more effectively with the Joint Staff and other services, and to establish clear parallels between different Air Force staffs. The Army and Navy have moved away from J-codes for their field staffs, but they have retained the organizing construct for their Pentagon staffs.

[2] Since maritime operations also take place over long distances, these missions might be appropriate for naval units as well.

[3] Johnson, 2007.

ers are likely to run the operation themselves, and they are unlikely to form a JTF.

A related question is how many different types of JTF headquarters does the Air Force need. The range of military operations is wide, and it is impossible to prepare for everything. JFCOM has responded to this problem by establishing separate templates for major combat, stability operations, and disaster relief. As air forces are not likely to play a leading role in most stability operations, it probably makes sense for the Air Force to focus on building JTF headquarters for disaster relief and for combat.[4] Since the Air Force does not usually lead JTFs, it might be best for the service to show first that it can effectively lead large and complex humanitarian operations so that, in time, it can develop more capability and build more of a reputation for competence that will lead to a JTF assignment dealing with combat.

There are some who advocate that the Air Force should establish unit type codes (UTCs) for specific types of JTF headquarters. UTCs categorize types of organizations into different classes of characteristics.[5] The idea is that the Air Force could create, for example, a list of UTCs for a JTF headquarters for a humanitarian operation and a list for a combat operation. Advocates for such an approach argue that taking this step would help the Air Force build the capability to lead. Others are more circumspect, arguing that the Air Force has a long way to go to reach true JTF headquarters capability, and that much work needs to be done before establishing UTCs.

Determine How Many JTF-Capable NAFs the Air Force Needs

Does every NAF need to be capable of leading a JTF? Organizations such as the Twentieth or the Twenty-Second Air Forces are unlikely

---

[4] Here we classify NFZ enforcement as a combat operation, not a stability operation.

[5] Kenneth Hill, "Force Packaging," IP-4200, Contingency Wartime Planning Course (CWPC), Maxwell Air Force Base, Ala.: Air University, College of Aerospace Doctrine, Research, and Education.

to be called upon to lead a JTF HQ.[6] Most of the NAFs that support functional combatant commands are also unlikely to be asked to serve as a JTF headquarters core.[7] The most likely candidates for JTF duty are NAFs that support regional combatant commands.

Each regionally oriented C-NAF deals with a different set of operational problems. Thirteenth Air Force might be called upon to lead a JTF for either combat or humanitarian operations in the Pacific theater. This monograph details how Third Air Force has had experience leading a humanitarian operation in Africa. It could well be called upon to lead a humanitarian operation in the EUCOM AOR. Seventeenth Air Force could be asked to lead a humanitarian operation or NEO in Africa for the newly constituted Africa Command (AFRICOM). This monograph also discusses how Ninth Air Force led a combat JTF in the Persian Gulf. It might play a similar role in protecting Iraq from future threats or in another conflict in that volatile region. Twelfth Air Force could conceivably be called upon to lead either a humanitarian or a combat operation in South America. NAFs need to engage in dialogue with their respective combatant commands about what sorts of operations they could conceivably be called upon to lead, and they need to work with their combatant commands to prepare for the most likely and/or most critical eventualities.[8]

Since resources are limited, the Air Force might want to refrain from developing JTF capability for each of its regionally oriented NAFs. If it is necessary to prioritize, the Air Force should first work to establish Third Air Force, Ninth Air Force, and Thirteenth Air Force as JTF capable. As a secondary effort, it should then turn to Twelfth Air Force and Seventeenth Air Force.

---

[6] Twentieth Air Force is responsible for intercontinental missiles. Twenty-Second Air Force provides training for reservist lift and support units.

[7] The most obvious exception to this is the Air Force Special Operations Command.

[8] On a lesser scale, Eleventh Air Force could conceivably lead a JTF responding to a humanitarian crisis in Alaska.

Determine How the Air Force Would Simultaneously Provide C-NAF and JTF Headquarters

If an Air Force entity is asked to lead a JTF, it is likely that air, space, and/or cyberspace forces will be expected to play an important role in the operation. Thus, in addition to the JTF headquarters, the Air Force will be expected to provide an air component to the operation. Within the C-NAF, either the AOC or the AFFOR staff could provide the basis for the JTF HQ. The advantage of using the AOC would be that it is postured to command operations, while the AFFOR staff is focused on management tasks. However, for cases in which an Air Force unit is asked to lead a JTF, the AOC would likely play a critical role in orchestrating joint air operations. The Air Force has made considerable efforts toward building the AOC into a competent organization that can plan and oversee air and space operations across a theater. Raiding the AOC to staff the JTF headquarters might endanger the AOC's effectiveness. The AFFOR staff would also have an important role to play in supporting Air Force forces in theater, but it is probably easier to replicate the functions of the AFFOR than those of the AOC.

If AFFOR staffs are expected to step up to the JTF headquarters role, the Air Force needs to consider how it will provide AFFOR functions. AFFOR staffs from other theaters could conceivably be called in to fill this role, but they and other likely augmentees would lack the regional expertise that would be necessary for them to function effectively.

Other than turning to an AFFOR staff to fill the JTF headquarters core, the Air Force might have other options, such as a wing staff for a small operation. Unfortunately, wing staffs, which focus on scheduling flights, are even less suited to the JTF role than are AFFOR staffs. It could also create a standing JTF headquarters unit that could deploy worldwide as needed. A worldwide JTF headquarters would lack regional-specific knowledge, and it would be difficult to maintain in the absence of operations.

Determine How the Air Force Would Man the Bulk of JTF Headquarters Positions

In addition to providing the JTF headquarters core, Air Force–led JTFs will need to augment the headquarters to reach its required strength. The Air Force might consider designating reserve or guard units as JTF headquarters augmentees.[9] Taking this step would allow the units to prepare in advance for their roles. The Air Force also might consider transferring staff from AFFORs from other theaters. Another option would be to find augmentees from MAJCOM and other Air Force staffs.

Determine How the JTF Headquarters Would Incorporate Other Services and Non-DoD Partners

In addition to its own augmentees, the Air Force will want personnel from other services, other government agencies, and coalition partners to round out the JTF headquarters staff. Augmentees who are not from the Air Force would provide a source of knowledge on force capabilities from other mediums, different perspectives on the nature of the operation and the best way to carry it out, and how to manage relationships with key actors in the theater. In addition, non–Air Force augmentees would ease the staffing burden for the Air Force. As we have mentioned, the JMD process is notoriously slow, and other options offer only partial remedies.

The Air Force also needs to consider how it will use personnel from other services. Traditionally, the JTF commander's deputy comes from another service. Other than this, there is little precedence for how to incorporate joint augmentees into a JTF headquarters. To the extent that an operation relies on the use of ground- or sea-based forces, the Air Force would obviously want to include Army, Navy, and Marine Corps personnel in roles that would allow the effective integration of their forces in the operation. Army and Marine Corps officers, who

---

[9] The Air Force has already done this on an individual basis.

have a reputation for competence in planning and in drafting orders, would be useful in these roles. Personnel with other useful skills could also contribute significantly to the success of the JTF headquarters if incorporated correctly. In addition, effective liaison with non–Air Force representatives would be important to increase JTF effectiveness and to maintain unity among the different organizations involved in JTF operations.

CHAPTER SEVEN
Recommendations

From our discussion of the theory and practice of JTFs, command-related developments in DoD, and four JTFs, we have developed a set of requirements for JTF headquarters and a list of issues for the Air Force to consider. We conclude this monograph by offering some recommendations for how the Air Force can increase its ability to field JTF headquarters cores. We have divided these recommendations into three basic categories: systems, people, and processes.

Systems

The bulk of the effort in building JTF headquarters capability involves people and processes, not technology. The technology necessary for these organizations exists and is relatively easy to procure. It is more difficult to provide qualified people, well-structured organizations, and sound processes. Nevertheless, having the proper systems is a necessary condition for a successful command and control entity.

Acquire Necessary Systems

To build effective JTF headquarters, the Air Force will need to upgrade selected AFFOR command centers. Some NAFs do not have standing AFFOR command centers, although they can stand them up if needed.[1] To help the AFFOR take on the role of a JTF core, AFFOR

---

[1] Personal communication with an AFFOR Chief of Staff, March 10, 2008.

command centers would need to have systems capable of sending and receiving information to and from fielded forces in the air, in space, at sea, and on land. JFCOM is currently working on creating "turnkey" capability, which is essentially a list of required systems for JTF headquarters operations. The Air Force would do well to work closely with JFCOM as it develops this program.

Determine the Desired Approach Toward Reach Back
As we discussed above, the Air Force is building an operations support facility at Langley Air Force Base to provide continuity of operations, data storage, and active support to fielded C-NAFs. As part of its effort to build JTF headquarters, the Air Force should consider how much these organizations will need to reach back to the OSF for their operations, and it should ensure that the OSF will be able to support JTF headquarters functions in addition to its performing AOC work. The Air Force should also consider how much reach back is necessary, possible, and desirable. Having command elements at remote locations outside of a theater lowers vulnerability and reduces costs. There is a tradeoff, however, in that there are questions about whether distributed command and control entities can be as effective as those that feature colocated personnel.

People

Reward Officers' Deep Experience with Joint, Interagency, and International Partners
Another step the Air Force should take is to reward officers with extensive experience working with other services, government agencies, and likely coalition militaries. At present, spending more than one tour with another organization is perceived by many as harmful to an officer's Air Force career. Further study would be necessary to determine whether this perception is grounded in fact. The Goldwater-Nichols Act does call for officers who have served in joint duty positions to be promoted at rates no less than other officers. A new points system will allow officers to nominate themselves for joint credit for serving with

other services, countries, and agencies.[2] Nevertheless, there is no provision within the Air Force personnel system to encourage officers to serve more than one tour outside the Air Force.

By ensuring that officers who have spent more than one tour with other organizations are promoted at a rate equal to or above that of others, the Air Force can send a message that it is interested in well-rounded officers who have gained specific knowledge about military operations in other domains, about how other organizations work, and more-general lessons about how to establish effective working relationships with non–Air Force personnel. Air Force officers who have spent considerable time with ground services could help ensure that an Air Force–led JTF headquarters properly incorporates the capabilities and perspectives of ground forces into war plans. Air Force officers who possess deep knowledge about the inner workings of other government agencies and coalition partners would also help coordinate efforts between an Air Force–led JTF headquarters and non–Air Force organizations. This sort of background is essential to the effective functioning of a JTF headquarters.

Reorient Professional Military Education

Air Force professional military education can help build the service's JTF capability. At present, Air Force–sponsored intermediate-level military education at Air Command and Staff College is geared toward providing a rigorous, broad-based education to Air Force officers.[3] This provides a good foundation for strategists and for the need to think flexibly and critically under conditions of chaos. It is important to have military officers who can think broadly and who can ensure that the actions of fielded forces meet the needs of policymakers. However, this approach does not lend itself well to more-functional planning associated with military operations. Strategy requires a broader perspec-

---

[2] Fred W. Baker III, "Officers Get New Joint Credit Qualification System," *American Forces Press Service*, July 30, 2007; and U.S. Department of Defense, *Department of Defense Strategic Plan for Joint Officer Management and Joint Professional Military Education*, Washington, D.C., April 3, 2006b.

[3] Interview with Air Command and Staff College official, July 26, 2007.

tive, akin to focusing on the forest, while planning requires a narrower focus, more like focusing on individual trees.

Again we can compare the Air Force to the Army, the service that usually provides JTF headquarters. Students at the School of Advanced Military Science (SAMS), the Army's elite second-year program, are expected to be staff planners. They are indoctrinated in the Military Decision Making Process (MDMP). MDMP offers a formal step-by-step method of gathering information; developing alternative courses of action; and rehearsing, executing, and assessing a military operation. SAMS graduates are sent to staff posts to be planners, and they are highly valued throughout the Army.

Like their Army counterparts, graduates of the School of Advanced Air and Space Studies (SAASS) are held in high regard by their peers. Unlike their Army counterparts, SAASS students are not normally sent to numbered units as planners. Instead of being trained to be planners, SAASS students are schooled to be strategists. The SAASS mission statement is as follows: "Produce *strategists* through advanced education in the art and science of air, space, and cyberspace power to defend the United States and protect its interests."[4]

To produce high-quality staffs and commanders of JTF headquarters, Air Force professional military education needs to produce both strategists and planners. Air Force personnel assignments and promotion practices should also reward planning ability as much as strategic insight.

Assign Competitive Personnel to AFFOR Staffs

One problem with using the AFFOR staff to man JTFs is the perception that AFFOR staffs are not typically populated with officers most competitive for promotion. Reports of below-primary-zone (BPZ) promotion rates, squadron commander experience, and intermediate and senior education completion for one NAF headquarters assigned to

---

[4] School of Advanced Air and Space Studies (SAASS) home page (emphasis added). Stephen Chiabotti, vice commandant of SAASS, writes that the SAASS mission "was narrowly defined to . . . produce strategists—not leaders, not warriors, not even planners" (Stephen D. Chiabotti, "A Deeper Shade of Blue: The School of Advanced Air and Space Studies," *Joint Forces Quarterly*, No. 49, 2nd quarter 2008).

support a regional combatant command indicate that personnel at the headquarters are considerably below the Air Force average for almost every indicator. This is reflected in Figure 7.1.

At present, many officers believe that to progress further in their careers, it helps to demonstrate technical proficiency in their field and to lead at the squadron, group, and wing levels. Staff tours are generally not as valued, unless they provide joint credit or offer the prospect of close association with an influential general officer. Tours at AFFOR staffs, in particular, are avoided by officers with high potential. If it is true that less-competitive officers are assigned to AFFOR staffs, this practice should be discontinued. If it is not true, then Air Force leaders should take steps to reverse the negative opinion of AFFOR service that is prevalent among Air Force personnel.

**Figure 7.1
Comparison of a Notional AFFOR Staff with Other Typical Air Force Staffs**

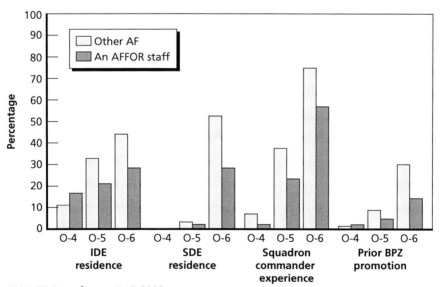

SOURCE: Data from a NAF, 2008.
NOTE: The source is purposely ambiguous to protect the anonymity of the staff.
RAND MG777-7.1

Train AFFOR Staffs

Current Air Force efforts to train C-NAF staff focus almost exclusively on the AOC. There is a week-long COMAFFOR Senior Staff Course (CSSC). Unfortunately, the audience is restricted mostly to colonel-level division chiefs, excluding most AFFOR staff members. The course is also limited to introducing different key actors and processes, such as the role of the Director of Space Forces (DIRSPACEFOR) and the Request for Forces (RFF) process.[5] For the rest of AFFOR staff, there is an online course offered through the College of Aerospace Doctrine, Research, and Education (CADRE), but this does not seem to indicate much of a commitment by the Air Force to prepare people to work on AFFOR staffs.

Above we noted that the Army trains staffs through its Battle Command Staff Training Program, and the Marine Corps does the same through its MAGTF Staff Training Program. The Air Force does have an Operational Command Training Program, but the group focuses on training for positions in the AOC. It does have one AFFOR trainer, but AFFOR training, even if requested, is rather broad in scope.[6]

Overall, there is evidence that the Air Force does not give as much value to staff work as to other experiences. Excellence in staff work is not considered to be as valuable for an officer as demonstrating technical proficiency in an aircraft or another professional area. The idea seems to be that Air Force officers are capable of learning how to do staff work on the job and that they do not need much training beforehand in the functions and processes of the AFFOR staff. This raises the question of whether AFFOR staffs are capable of better work than the work they do at present. More germane to our study, one can make the case that a lack of training makes an AFFOR staff a less attractive candidate to form a JTF headquarters core than its Army equivalent. Building a credible training program for AFFOR staffs both for their AFFOR roles and for their ability to transform into JTF cores is a

---

[5] USAF Expeditionary Center, "COMAFFOR Senior Staff Course Schedule," Fort Dix, N.J., April 2007.

[6] Interview with Lt Col Donald Finley, Operational Command Training Program Squadron Commander, August 8, 2007.

significant undertaking, but one that the Air Force must embrace to benefit the joint force.

Processes

In addition to changes in the way that the Air Force educates, trains, and evaluates its people, the Air Force could make a number of process changes to boost its capability to lead JTFs.

Designate JTF-Capable Organizations

Air Force leadership should task formally selected C-NAFs to be capable of forming the core of a JTF. In documents such as the C-NAF Program Action Directive (PAD), the AFFOR Architecture, and the command and control enabling concept, the Air Force mandates that C-NAFs should be capable of forming JTF headquarters cores, but none of these documents specify which organizations should be capable of commanding what sort of operations or how they would prepare for such a role.

In consultation with the designated C-NAFs and their respective combatant commands, the Air Force should specify general mission areas that the C-NAF should be capable of undertaking. Such a step would require the Air Force to consider exactly which units need to be capable of which missions. It would also require a consideration of the resources necessary to prepare C-NAFs for the JTF headquarters role. The process of crafting a policy statement designating JTF-capable units would itself help the Air Force begin to build the capability.

Use Exercise Programs

Once designated as JTF-capable organizations by the Air Force, C-NAFs will need to develop that capability and ensure their readiness for action. They can do this by planning and participating in exercises, either on their own or as part of a larger exercise with their combatant commands.[7] Third Air Force's work in the Flexible Leader exercise serves as a good example. Preparing for the JTF headquarters role helps the organization increase readiness and also demonstrates to combatant commanders and other decisionmakers that the organization is capable of taking on such a role.

Place More Emphasis on Planning

As we mentioned above, the Air Force tends to focus more on producing strategists than planners. In general, the service has a reputation for not taking planning as seriously as other services. To some extent, this is a natural result of the nature of the Air Force as an institution and a reflection of the way the service fights. The fighting units of the Air Force are college-educated officers. They are highly trained and operate extremely expensive aircraft. They operate far from their home bases and have considerable latitude in how to accomplish their missions. Accordingly, the Air Force places value on the ability to improvise and to respond to changing circumstances.[8] As a result, Air Force officers often express reluctance to submit to systematic processes, such as MDMP.

The Air Force does have its own planning process for air operations, the joint air and space estimate process, which produces the joint air and space operations plan (JAOP).[9] It teaches this process as part of the AOC training conducted by the 505th Training Squadron at Hurlburt Air Force Base. It is important to note, however, that the Air Force views JAOP as the "vehicle through which the JFACC directs joint

---

[7] While it may be necessary to institute new exercises, the current glut of exercise programs suggests that it would be desirable to make better use of currently existing events.

[8] For more on the Air Force's vision of itself, as well as the other services' visions of themselves, see Carl H. Builder, *The Masks of War: American Military Styles in Strategy and Analysis,* Baltimore, Md.: Johns Hopkins University Press, 1989.

[9] JAOP is the basic description of how the JFACC will integrate forces from different services and countries into an air campaign. It identifies the JFACC's intent, sets objectives, and discusses rules of engagement. It lays out command arrangements, ISR collection priorities, and support requirements. Headquarters, U.S. Air Force, 2007b, pp. 100–101, 113; Headquarters, U.S. Air Force, *Air Warfare: Air Force Doctrine Document 2-1,* Washington, D.C., January 22, 2000, pp. 38–45.

aerospace power."¹⁰ It is primarily a component-level process, and it is not intended to be used to direct forces in other mediums. Moreover, it is not considered to be as rigorous as MDMP.

To address this gap, the Air Force has several options. It could adapt the JAOP process to address the use of forces from other mediums. Alternatively, it could adopt a process similar to the Army's MDMP. Third, it could create a new planning process that combines the rigor of MDMP with the comparative advantages associated with air power.¹¹ Regardless of which approach it takes, it should teach that approach at Air Command and Staff College and other venues to create a cadre of planners.

Write a Directive on Air Force JTF Operations

One step that would help the Air Force build JTF capability would be to draft a directive on how the Air Force would prepare for and execute the JTF headquarters role. Similar to the designation of JTF-capable organizations, the drafting of this document would require discussions across the Air Force about the best way to build and use an Air Force JTF headquarters. The directive would need to lay out how the Air Force as an institution and how individual AFFOR staffs would build JTF headquarters capability, outlining steps discussed here and tasking different Air Force entities to help make the vision of an Air Force JTF headquarters a reality.

The Air Force might also need to draft CONOPS for how an Air Force JTF headquarters might operate. This document would take into consideration how the Air Force's perspective on military operations would be translated into the operation of a JTF headquarters. It would discuss how an Air Force–led JTF headquarters would be organized, how it would prepare for operations, and how it would define equipment requirements.

---

¹⁰ Headquarters, U.S. Air Force, 2000, p. 38.

¹¹ For an argument for creating a universal planning process for U.S. armed forces, see COL Joseph Anderson, USA, and COL Nathan K. Slate, USA, "The Case for a Joint Military Decisionmaking Process," *Military Review,* September–October 2003.

Learn JTF Headquarters Processes

There are a number of processes associated with the operation of JTF headquarters that Air Force personnel would need to master. For example, it will be necessary to understand how to request headquarters staff through the various means discussed above, such as the JMD process. Some of this work will involve filing formal requests, but it will also require knowledge of how personnel allocation processes in the joint community work. It will also require the Air Force to understand what sorts of people it needs to ask for and what roles they will play in the JTF headquarters. Similarly, the Air Force will need to know how to request forces to employ in the operation itself.

In addition, the Air Force will need to have the capability to issue formal orders. At the component level, commanding through email and PowerPoint might work, but at the JTF level, where it is necessary to communicate clearly with individuals from other services and countries that operate in different mediums, formal orders will be necessary. Anecdotal evidence indicates that there are surprisingly few Air Force officers who have experience drafting formal orders. To communicate effectively with other services, the Air Force will need to change.

Create a Capability to Deploy Headquarters

In general, Air Force command centers do not deploy. Instead, the Air Force's operational units deploy or operate from home bases, often traveling hundreds if not thousands of miles. Since modern communications allow real-time connectivity at long distances, the AOC does not have to be near the actual AOR to provide command and control. In contrast, the Army's vision is that the command center deploys along with the troops. The Army's vision compares well to the QDR tasking, which foresees "[r]apidly deployable" JTF headquarters.[12] The Air Force could engage in debate over whether JTF headquarters need to be near the area of operations. After all, CENTCOM commanded much of the early stages of Operation Enduring Freedom from Tampa, Florida. However, there are advantages that come with deploying the

---

[12] U.S. Department of Defense, 2006a, p. 59.

JTF headquarters, such as better access to information and the ability to meet face to face with coalition partners located in the AOR.

The Air Force does seem to recognize the advantages of deploying command centers. The Air Force has called for the development of a "flexible-response" requirement that would allow it to deploy elements of selected AOCs.[13] Gaps in the development of deployment capability, and the fact that it has not yet been extended to the AFFOR staff, appear to be the results more of resource limitations and a practical decision to pursue other command-related initiatives than a philosophical opposition to deployability.[14]

Create a Champion for Air Force Command
The last recommendation we make has more to do with the Air Force as an institution than with the operation of JTF headquarters. The Air Force should designate one office to develop policy and identify resource requirements for Air Force command-and-control entities. Part of the problem with developing Air Force command capabilities lies in the lack of a single designated advocate in this critical area. Multiple offices play a role in obtaining resources and setting policy for Air Force command centers, including the Air Staff, the Air Combat Command, and the C-NAFs themselves. Weapon systems such as aircraft have program managers who enjoy the authority and bureaucratic muscle that they employ on behalf of their platforms. There is no analog to this for Air Force command, even though the AOC has been designated a weapon system.[15] Budgets for aircraft are represented in the Air Force's Program Objective Memoranda. The command community needs similar representation.

---

[13] Headquarters, U.S. Air Force, 2006c, p. 16.

[14] We are grateful to Gilbert Braun for discussion on this point.

[15] Gen Michael Ryan designated the AOC a weapon system on September 8, 2000. U.S. Air Force, Global Cyberspace Integration Center, "History of the Global Cyberspace Information Center," Langley Air Force Base, Va., updated January 2008.

Conclusion

If the Air Force takes the steps necessary to build and maintain JTF-capable units, both the service and the nation would benefit. An Air Force–led JTF headquarters capability would assist the U.S. Armed Forces by providing another option, a different type of JTF leadership than the one usually provided for by the Army or another service. The investment is not trivial, but the rewards are great.

APPENDIX A
Joint Task Forces Since 1990

Table A.1
Joint Task Forces Since 1990

Joint Task Force	Mission/Operation	Start	End	Service Command	Rank
JTF-Bravo	Theater Security Cooperation (Honduras)	1983	Present	U.S. Army	06
JTF-6	Counternarcotics	1989	Present	U.S. Army	07
JTF–Sharp Edge	NEO (Liberia)	1990	1991	USN	06
JTF–Proven Force	**Combat air operations (Iraq)**	**1990**	**1991**	**USAF**	**08**
JTF–Patriot Defender	Missile defense (Israel)	1991	1991	U.S. Army	06
JTF–Fiery Vigil	**Humanitarian evacuation (PI)**	**1991**	**1991**	**USAF**	**08**
JTF–Quick Lift	**NEO (Zaire)**	**1991**	**1991**	**USAF**	**07**
JTF–Guantanamo Haitian Refugees	HA/refugee support	1991	1993	USMC	07
JTF–Provide Comfort	HA (Northern Iraq)	1991	1991	U.S. Army	09
JTF–Sea Angel	HA (Bangladesh)	1991	1991	USMC	08
JTF–Provide Hope	**HA (Soviet Union)**	**1992**	**1993**	**USAF**	**06**
JTF–Provide Transition	**Support to Peace Support Operation (PSO) (Angola)**	**1992**	**1992**	**USAF**	**05**
JTF–Los Angeles	Military support to civil authorities	1992	1992	U.S. Army	08

93

Table A.1—Continued

Joint Task Force	Mission/Operation	Start	End	Service Command	Rank
JTF–SWA (Southern Watch)	NFZ enforcement (Iraq)	1992	2003	USAF	08
JTF-Miami	HA/Hurricane Andrew	1992	1992	U.S. Army	09
JTF–Typhoon Omar	HA/Guam ("Marianas")	1992	1992	USN	08
JTF-Iniki	HA/Hawaii	1992	1992	U.S. Army	09
JTF–Restore Hope	HA (Somalia)	1992	1993	USMC	09
JTF–Provide Relief	HA (Somalia/Kenya)	1992	1993	USMC	07
JTF–Provide Promise	PSO (Former Yugoslavia)	1992	1996	USN	10
JTF-Somalia	Support to UN operations	1993	1994	U.S. Army	08
JTF–Provide Refuge	HA/refugee support (Kwajalein Atoll)	1993	1993	U.S. Army	08
JTF–Deny Flight/ Decisive Edge	NFZ enforcement (Bosnia/ Herzegovina)	1993	1995	USAF	08
JTF-Haiti/JTF-120	PSO	1993	1995	USN	07
JTF-160	Sea Signal (refugee support)	1993	1996	Rotational	07
JTF–Support Hope/ Quiet Resolve	HA (Rwanda)	1994	1994	U.S. Army	07
JTF–Deliberate Force	Air campaign against Serbian military forces	1995	1995	USAF	09
JTF–Assured Response (which supported JTF–Quick Response)	NEO/Embassy security (Liberia)/JTF–Quick Response: NEO in Central African Republic	1996	1996	USMC	06
JTF–Guardian Assistance	HA (Great Lakes Region)	1996	1996	U.S. Army	08
JTF–Pacific Haven	Refugee screening (Guam)	1996	1997	USAF	08
CTF–Northern Watch	NFZ enforcement	1997	2003	USAF	07
JTF–Assured Lift	NEO/Movements (Liberia)	1997	1997	USAF	08

Table A.1—Continued

Joint Task Force	Mission/Operation	Start	End	Service Command	Rank
JTF–Guardian Retrieval	NEO (Congo)	1997	1997	U.S. Army	06
JTF–Noble Obelisk	NEO (Sierra Leone)	1997	1997	USMC	06
JTF–Eagle Vista	**Support to U.S. President's visit to Africa**	**1998**	**1998**	**USAF**	**08**
C/JTF-Kuwait	FDO/MCO	1998	2003	U.S. Army	09
JTF–Shining Presence	Missile defense (Israel)	1998	1999	U.S. Army	08
JTF–Noble Anvil	Air operations in support NATO OAF in Kosovo	1999	1999	USN	10
JTF–Shining Hope	**HA (Kosovo)**	**1999**	**1999**	**USAF**	**08**
JTF-AH	Allied Harbor	1999	1999	U.S. Army	10
JTF–Skilled Anvil	Peace Enforcement planning for Kosovo	1999	2000	U.S. Army	10
JTF–Civil Support	Summit Guard	1999	Present	U.S. Army	08
JTF-Falcon	Joint Guardian (Kosovo)	1999	2004	U.S. Army	07
JTF–Atlas Response/Silent Promise	**Humanitarian relief (Mozambique)**	**2000**	**2000**	**USAF**	**08**
JTF-O/JTF-509	Winter Olympics (National Special Security Event [NSSE])	2001	2002	U.S. Army	07
International Security and Assistance Force (ISAF)	Enduring Freedom (Afghanistan)	2001	Present	U.S. Army	10
Combined Force Command–Afghanistan/ Combined Security Transition Command–Afghanistan (CSTC-A)	Enduring Freedom (Afghanistan)	2001	Present	U.S. Army	08

Table A.1—Continued

Joint Task Force	Mission/Operation	Start	End	Service Command	Rank
Combined Joint Task Force (CJTF)-180, -76, -82, -101	Enduring Freedom (Afghanistan)	2001	Present	U.S. Army	08
JTF-Piton	St. Lucia	2001	2001	Unknown	
JTF–Homeland Defense	Defense Support for Civilian Authorities for Hawaii	2001	Present	U.S. Army	09
JTF-510	**Enduring Freedom (Philippines)**	**2002**	**2002**	**SOCPAC**	**07**
JTF-PAKLNO	CENTCOM LNO to Pakistan	2002	Present	U.S. Army	06
JTF-160	Enduring Freedom (Detainees)	2002	2002	U.S. Army	07
JTF-170	Enduring Freedom (Detainees)	2002	2002	U.S. Army	08
JTF-GTMO	Enduring Freedom (Detainees)	2002	Present	USN	07
JTF–Autumn Return	**NEO in Cote d'Ivoire**	**2002**	**2002**	**USAF**	**07**
JTF-519	U.S. PACOM SJTF	2002	Present	USN	10
JTF-HOA	Enduring Freedom (Horn of Africa)	2002	Present	USN	08
CJTF-HOA	Enduring Freedom (HOA)	2002	Present	USN	07
JTF-H	Secure Tomorrow (stability and support operation [SASO])	2002	2002	USMC	07
JSOTF-P	**Global War on Terror**	**2002**	**Present**	**SOCPAC**	**06**
OMA-A/OSC-A	Enduring Freedom-Afghanistan	2002	Present	U.S. Army	07
JTF-4	CENTCOM	2003	2003	Unknown	
JTF-TF-121	**Enduring Freedom/Iraqi Freedom**	**2003?**	**Present**	**SOCCENT**	**07**
JTF-I	Sheltering Sky (NEO)	2003	2003	U.S. Army	08

Table A.1—Continued

Joint Task Force	Mission/Operation	Start	End	Service Command	Rank
JTF-Liberia	SASO	2003	2003	U.S. Army	08
JTF-58	NSSE: security for U.S. President's visit	2003	2003	USN	10
JTF-AS	Aztec Silence (Global War on Terror)	2003	Present	USN	10
CJTF-7	Iraqi Freedom	2003	2004	U.S. Army	09
JTF-JSOTF-AP	Iraqi Freedom	2003?	Present	U.S. Army	06
MNC-I	Iraqi Freedom	2004	Present	U.S. Army	09
CJTF-Phoenix	Enduring Freedom	2004	Present	U.S. Army	07
MNF-I	Iraqi Freedom	2004	2004	U.S. Army	10
CJTF-Haiti	Haiti (SASO)	2004	2005	USMC	07
MNSTC-Iraq	Iraqi Freedom (Train and Equip)	2004	Present	U.S. Army	09
JTF-134	Iraqi Freedom (Detainees)	2004	Present	U.S. Army	08
CJTF/CSF-536	Unified Assistance (Tsunami relief)	2004	2005	USMC	09
JTF-GRD	Iraqi Freedom (Corps of Engineers: Gulf Region Division)	2004	Present	U.S. Army	07
JTF-AFIC	Armed Forces Inauguration Committee	2004	2005	U.S. Army	08
JTF-NSJ	Boy Scouts of America Jamboree (National Scout Jamboree)	2005	2005	U.S. Army	08
JTF-Paladin	Iraqi Freedom	2005?	Present	U.S. Army	06
(C)JSOTF-HOA/ SOCCENT-HOA	Enduring Freedom (HOA), Special Operations Command Central	2005	Present	U.S. Army	06
JTF-510	PACOM crisis response	Unk.	Unk.	Rotational	07

Table A.1—Continued

Joint Task Force	Mission/Operation	Start	End	Service Command	Rank
JTF–Joint Area Support Group (JASG)	Iraqi Freedom	2005	Present	U.S. Army	06
Joint Contracting Command–Iraq/Afghanistan (JCC-I/A)	Enduring Freedom/Iraqi Freedom	2005	Present	Rotational	08
IAG	Iraqi Freedom (Iraq Assistance Group)	2005	Present	U.S. Army	07
CJTF-Troy	Iraqi Freedom	2005	Present	U.S. Army	06
JTF-JTTR	Enduring Freedom/Iraqi Freedom (Joint Theater Trauma Registry)	2005	Present	U.S. Army	06
JTF-ITFC	Iraqi Freedom (Iraq Threat Finance Cell)	2006	Present	U.S. Army	05
JTF-South	Operation Enduring Freedom	Unk.	Present	U.S. Army	08

SOURCE: Various sources, including E. W. Cobble, H. H. Gaffney, and D. Gorenburg, *For the Record: All U.S. Forces Responses to Situations, 1970–2000* (with addition covering 2000–2003), Alexandria, Va.: Center for Naval Analyses, Report CIM D0008414.A2/final, February 2002; George Steward, Scott M. Fabbri, and Adam B. Siegel, *JTF Operations Since 1983,* Alexandria, Va.: Center for Naval Analyses, 1994.

NOTES: Bold indicates Air Force–led JTFs. Unk. = unknown.

APPENDIX B

Joint Manning Document Data from Selected Joint Task Forces

Table B.1
Joint Manning Document for Joint Task Force Atlas Response Headquarters

JHQ Element	Personnel	USAF	USN	USMC	U.S. Army	SOF	EUCOM/JAC
Commander	1	1					
CMD Group	12	11			1		
J-1	8	7			1		
J-2	24	22			2		
J-3	21	18	1	1	1		
J-4	23	19	4				
J-5	4	4					
J-6	17	14			1	2	
Medical	9	6	3				
Chaplain	3	3					
Legal	4	4					
Public Affairs Office/Joint Information Board	9	9					
JHQ Command	4	4					
Liasion officers	8	6				1	1
Total HQ	147	128	8	2	7	1	1
Percentage of HQ by service		87%	5%	1%	5%	1%	1%

SOURCE: Atlas Response, *Joint Manning Document*, not dated. Quoted in unpublished research by Armando X. Estrada and Janice H. Lawrence, on Joint Task Force operations, 1990–2004, at the Graduate School of International Business and Public Policy, Monterey, Calif.

Table B.2
JTF/CSF—Unified Assistance (CSF-536) Headquarters

JHQ Element	Personnel	USAF	USN	U.S. Army	USMC	SOF	CIV
		152	85	12	653	83	1
Total HQ	986						
Percentage of HQ by service		15%	9%	1%	66%	8%	0%

SOURCE: JTV/CSF—Unified Assistance, *Joint Manning Document*, December 31, 2004.

NOTE: A formal JMD apparently was never developed and submitted to the Joint Staff because of time constraints; rather, only gross personnel numbers are available. Moreover, note that 458 of the 986 JTF HQ total personnel were located in theater.

Table B.3
Joint Task Force—Noble Anvil Headquarters

JHQ Element	Personnel	USAF	USN	U.S. Army	USMC	SOF	EUCOM/JAC
Commander	1	0	1	0	0	0	0
CMD Group	17	3	7	3	4	0	0
J-1	15	3	7	3	2	0	0
J-2	77	18	26	9	6	0	18
J-3	96	23	37	18	10	8	0
J-4	29	13	8	7	1	0	0
J-5	21	1	4	6	8	2	0
J-6	58	18	16	16	8	0	0
Medical	7	2	3	2	0	0	0
IG	1	0	1	0	0	0	0
Chaplain	1	0	1	0	0	0	0
Legal	1	0	1	0	0	0	0
Public Affairs Office/ Joint Information Board	2	0	1	0	1	0	0
Total HQs	326	81	113	64	40	10	18
Percentage of HQ by service		25%	35%	20%	12%	3%	6%

SOURCE: JTF-NA, *Joint Manning Document*, not dated, document provided by Historian's Office, U.S. European Command, Stuttgart.

Table B.4
Joint Task Force—Southwest Asia Headquarters

JHQ Element	Personnel	USAF	USN	U.S. Army	USMC	SOF	CIV
		186	31	23	5	0	6
Total HQ	251						
Percentage of HQ by service		74%	12%	9%	2%	0%	2%

SOURCE: HQ JTF Southwest Asia, not dated. GlobalSecurity.org, "Joint Task Force Southwest Asia," April 26, 2005.

NOTE: The RAND team put considerable research into locating a detailed JMD document but was unable to find one.

Bibliography

Books, Reports, and Articles

Ackerman, Robert K., "New Flight Plan for Air Force Intelligence," *Signal Magazine,* March 2007.

Allard, Kenneth, *Command, Control and the Common Defense,* Washington, D.C.: National Defense University Press, revised edition, 1996.

Anderson, COL Joseph, USA, and COL Nathan K. Slate, USA, "The Case for a Joint Military Decisionmaking Process," *Military Review,* September–October 2003.

Baker, Fred W. III, "Officers Get New Joint Credit Qualification System," *American Forces Press Service,* July 30, 2007.

"Battle Alert in the Gulf," transcript of Nova, Public Broadcasting System, aired February 23, 1999. As of August 4, 2008:
http://www.pbs.org/wgbh/nova/transcripts/2608battlegroup.html

Belote, Lt Col Howard D., USAF, *Once in a Blue Moon: Airmen in Theater Command—Lauris Norstad, Albrecht Kesselring, and Their Relevance to the Twenty-First Century Air Force,* Maxwell Air Force Base, Ala.: Air University Press, CADRE Paper No. 7, July 2000.

Berkhouse, L., F. W. McMullan, S. E. Davis, and C. B. Jones, *Operation Crossroads—1946: United States Atmospheric Nuclear Weapons Tests, Nuclear Test Personnel Review,* Washington, D.C.: Defense Nuclear Agency, NTIS, DNA 6032F, 1984.

Bonds, Timothy, Myron Hura, and Thomas-Durell Young, *Enhancing Army Joint Force Headquarters Capabilities,* Santa Monica, Calif.: RAND Corporation, MG-675-A, forthcoming.

Builder, Carl H., *The Masks of War: American Military Styles in Strategy and Analysis,* Baltimore, Md.: Johns Hopkins University Press, 1989.

Builder, Carl H., Steven C. Bankes, and Richard Nordin, *Command Concepts: A Theory Derived from the Practice of Command and Control,* Santa Monica, Calif.: RAND Corporation, MR-775-OSD, 1999. As of October 14, 2008:
http://www.rand.org/pubs/monograph_reports/MR775/

Chiabotti, Stephen D., "A Deeper Shade of Blue: The School of Advanced Air and Space Studies," *Joint Forces Quarterly,* No. 49, 2nd quarter 2008.

Cobble, E. W., H. H. Gaffney, and D. Gorenburg, *For the Record: All U.S. Forces Responses to Situations, 1970–2000* (with addition covering 2000–2003), Alexandria, Va.: Center for Naval Analyses, Report CIM D0008414.A2/final, February 2002.

Cohen, Mike, "Mozambique-Floods," *Portsmouth Herald,* March 7, 2000.

Colaizzi, Jennifer, "Positive Review for Joint Manpower Exchange Program," *USJFCOM Public Affairs,* September 21, 2005. As of November 13, 2007:
http://www.jfcom.mil/newslink/storyarchive/2005/pa092105.htm

Correll, John T., "Northern Watch," *Air Force Magazine,* Vol. 83, No. 2, July 2000.

Daalder, Ivo H., and Michael E. O'Hanlon, *Winning Ugly: NATO's War to Save Kosovo,* Washington, D.C.: Brookings Institution Press, 2000.

Estrada, Armando X., *Joint Task Force Requirements Determination: A Review of the Organization and Structure of Joint Task Forces,* Monterey, Calif.: Naval Postgraduate School, Graduate School of Business and Public Policy, March 2005.

Garamone, Jim, "QDR Approves Joint Force Headquarters Concept," *American Forces Press Service,* October 29, 2001.

GlobalSecurity.org, "Joint Task Force–Southwest Asia," April 26, 2005.

Gordon, John IV, and Jerry Sollinger, "The Army's Dilemma," *Parameters,* Summer 2004.

Grant, Rebecca, "Marine Air in the Mainstream," *Air Force Magazine,* Vol. 87, No. 6, June 2004.

———, "Why Airmen Don't Command," *Air Force Magazine,* March 2008, pp. 46–49.

Graybar, Lloyd J., "The 1946 Atomic Bomb Tests: Atomic Diplomacy or Bureaucratic Infighting?" *The Journal of American History,* Vol. 72, No. 4, March 1986.

Hill, Kenneth, "Force Packaging," IP-4200, Contingency Wartime Planning Course (CWPC), Maxwell Air Force Base, Ala.: Air University, College of Aerospace Doctrine, Research, and Education. As of September 16, 2008:
http://www.fas.org/man/dod-101/usaf/docs/cwpc/4200-FO.htm

Hittle, Lt Col J. D., USMC, *The Military Staff: Its History and Development,* Harrisburg, Pa.: The Military Service Publishing Company, 1949.

Hoefel, CAPT J. Stephen, USN, *U.S. Joint Task Forces in the Kosovo Conflict,* Naval War College Paper, Newport, R.I.: U.S. Naval War College, May 16, 2000.

Hoehn, Andrew R., Adam Grissom, David A. Ochmanek, David A. Shlapak, and Alan J. Vick, *A New Division of Labor: Meeting America's Security Challenges Beyond Iraq,* Santa Monica, Calif.: RAND Corporation, MG-499-AF, 2007. As of October 14, 2008:
http://www.rand.org/pubs/monographs/MG499/

Hosmer, Stephen T., *The Conflict over Kosovo: Why Milosevic Decided to Settle When He Did,* Santa Monica, Calif.: RAND Corporation, MR-1351-AF, 2001. As of October 14, 2008:
http://www.rand.org/pubs/monograph_reports/MR1351/

Johnson, David E., *Learning Large Lessons: The Evolving Roles of Ground Power and Air Power in the Post–Cold War Era,* Santa Monica, Calif.: RAND Corporation, MG-405-1-AF, 2007. As of October 14, 2008:
http://www.rand.org/pubs/monographs/MG405-1/

Jumper, Gen John P., USAF, "Rapidly Deploying Aerospace Power: Lessons from Allied Force," *Aerospace Power Journal,* Winter 1999.

Kitfield, James, "The Long Deployment," *Air Force Magazine,* Vol. 83, No. 7, July 2000.

Koh, Col Mark, Singapore Armed Forces, "Operation Unified Assistance—A Singapore Liaison Officer's Perspective," *Liaison,* Vol. 3, No. 3, 2005.

Lambeth, Benjamin S., *The Transformation of American Air Power,* Ithaca, N.Y.: Cornell University Press, 2000.

———, *NATO's Air War for Kosovo: A Strategic and Operational Assessment,* Santa Monica, Calif.: RAND Corporation, MR-1365-AF, 2001. As of October 14, 2008:
http://www.rand.org/pubs/monograph_reports/MR1365/

———, *Air Power Against Terror: America's Conduct of Operation Enduring Freedom,* Santa Monica, Calif.: RAND Corporation, MG-166-1-CENTAF, 2005. As of October 14, 2008:
http://www.rand.org/pubs/monographs/MG166-1/

Lopez, SSgt C. Todd, USAF, "Changes Planned for ISR Community," *Air Force Print News,* January 30, 2007.

McBride, Hugh C., "New Plans and Operations Center Exemplifies EUCOM Transformation," *American Forces Press Service,* October 30, 2003.

Miles, Donna, "Core Elements Improve Crisis Response, Combat Ops," *American Forces Press Service,* March 23, 2006.

Mongillo, LCDR Nicholas, USN, *Navy Integration into the Air Force–Dominated JFACC,* Newport, R.I.: Naval War College, February 8, 2003.

Morabito, Maj. Antonio III, USMC, *NATO Command and Control: Bridging the Gap,* Newport, R.I.: Naval War College, February 5, 2001.

Muradian, Vago, "USAF Struggles with Budget Shortfall," *Defense News,* August 20, 2007.

Murphy, VADM Dan, USN, "The Navy in the Balkans," *Air Force Magazine,* Vol. 82, No. 12, December 1999.

Nardulli, Bruce, Walter L. Perry, Bruce R. Pirnie, John Gordon IV, and John G. McGinn, *Disjointed War: Military Operations in Kosovo, 1999,* Santa Monica, Calif.: RAND Corporation, MR-1406-A, 2002. As of October 14, 2008: http://www.rand.org/pubs/monograph_reports/MR1406/

Nelson, Michael A., and Douglas J. Katz, "Unity of Control: Joint Air Operations in the Gulf—Part Two," *Joint Forces Quarterly,* Summer 1994.

Plunkett, A. J., "U.S. Still Polices No-Fly Zone over Southern Iraq: Monotony Rules 4-Year Mission," *Newport News Daily Press,* December 25, 1994.

Priest, Dana, "The Battle Inside Headquarters; Tension Grew with the Divide over Strategy," *Washington Post,* September 21, 1999, p. A1.

Putrich, Gayle S., "USAF Reorganizing Intelligence Command," *Defense News,* January 30, 2007.

Schnabel, James F., *History of the Joint Chiefs of Staff,* Vol. 1, *The Joint Chiefs of Staff and National Policy, 1945–1947,* Washington, D.C.: Office of Joint History, Office of the Chairman of the Joint Chiefs of Staff, 1996.

Sligh, Robert, *ATLAS RESPONSE: Official History of Operation ATLAS RESPONSE,* Ramstein Air Base, Germany: Third Air Force, 2000.

Stavridis, ADM James, USN, and CAPT Robert Girrier, USN, *Watch Officer's Guide: A Handbook for All Deck Watch Officers,* 15th ed., Annapolis, Md.: Naval Institute Press, 2007.

Steward, George, Scott M. Fabbri, and Adam B. Siegel, *JTF Operations Since 1983,* Alexandria, Va.: Center for Naval Analyses, 1994.

Strickland, Lt Col Paul C., USAF, "USAF Aerospace-Power: Decisive or Coercive," *Aerospace Power Journal,* Fall 2000.

Tsunami Evaluation Coalition, *Joint Evaluation of the International Response to the Indian Ocean Tsunami: Synthesis Report,* London, U.K.: Active Learning Network for Accountability and Performance in Humanitarian Action, July 2006.

van Creveld, Martin, *Command in War,* Cambridge, Mass.: Harvard University Press, 1985.

Worden, Col Mike, USAF, *Rise of the Fighter Generals: The Problem of Air Force Leadership, 1945–1982,* Maxwell Air Force Base, Ala.: Air University Press, March 1998.

Worley, D. Robert, *Shaping U.S. Military Forces: Revolution or Relevance in a Post–Cold War World,* Westport, Conn.: Praeger Security International, 2006.

Zimmerman, COL Douglas K., USA, "Understanding the Standing Joint Force Headquarters," *Military Review,* July–August 2004.

U.S. Government Documents

Air Combat Command, "Air Force Forces Component Numbered Air Force Operations Support Facility Functional Concept," November 8, 2006.

Battle Command Training Program, "Battle Command Training Program: Commander's Overview Briefing," November 16, 2006.

BCC [blind carbon copy] to 3AF [3rd Air Force]/CCEA, "R031835Z Mar 00 EXORD for Operation ATLAS RESPONSE," email, March 2000.

Chairman of the Joint Chiefs of Staff, "Individual Augmentation Procedures," Instruction, Washington, D.C.: The Joint Staff, CJCSI 1301.01C, January 1, 2004.

Combined Arms Doctrine Directorate, "Command and Control in the Modular Force: Echelons and the Road to Modularity," July 3, 2007a.

———, "Army Organizations as Joint Capable Headquarters," briefing, August 13, 2007b.

Cossa, Ralph A., "South Asian Tsunami: U.S. Military Provides 'Logistical Backbone' for Relief Operation," *e-Journal USA, Foreign Policy Agenda,* November 2004, updated March 2005. As of November 13, 2007: http://usinfo.state.gov/journals/itps/1104/ijpe/cossa.htm

Defense Joint Intelligence Operations Center, "The Defense Joint Intelligence Operations Center: Overview," briefing, September 13, 2006.

Deptula, Lt Gen David A., *CSF-536 Joint Force Air Component Commander (JFACC)/Air Force Forces Commander (AFFOR) Lessons and Observations,* Pacific Air Forces, May 17, 2005.

Elder, Lt Gen Bob, "Air Force Cyberspace Command: Report to CSAF [Chief of Staff of the Air Force]," briefing March 23, 2007.

Ellis, ADM James O., USN, "A View from the Top," briefing, October 21, 1999.

Gilbert, Col S. Taco, "DIRMOBFOR [Director of Mobility Forces] JTF ATLAS RESPONSE: After Action Report," February 28, 2000.

Headquarters, Pacific Air Forces (PACAF), *With Compassion and Hope: The Story of Operation Unified Assistance—Air Force Support for Tsunami Relief Operations in Southeast Asia 25 December 2004–15 February 2005,* Hickam Air Force Base, Hawaii: PACAF, Air Education and Training Command's Historian office, not dated.

Headquarters, Standing Joint Force, "Improving Readiness for Joint Task Force Headquarters: Concept of Operations," unpublished Joint Force Command research, July 13, 2007.

Headquarters, U.S. Air Force, *Air Warfare: Air Force Doctrine Document 2-1,* Washington, D.C., January 22, 2000.

———, *Air Force Basic Doctrine: Air Force Doctrine Document 1,* Washington, D.C., November 17, 2003.

———, "Air Force Forces Command and Control Enabling Concept," Washington, D.C., May 25, 2006a.

———, *Implementation of the Chief of Staff of the Air Force Direction to Establish an Air Force Component Organization,* Program Action Directive 06-09, September 15, 2006b.

———, *Air and Space Operations Center Crew Roadmap,* Washington, D.C., September 20, 2006c.

———, "USAF Intelligence Way Ahead," briefing, Washington, D.C., January 16, 2007a.

———, *Operations and Organization, Air Force Doctrine Document 2,* Washington, D.C., April 3, 2007b.

Joint Publication 1-02, *The Department of Defense Dictionary of Military and Associated Terms,* Washington, D.C.: The Joint Staff, 2007.

Joint Publication 3-0, *Joint Operations,* Washington, D.C.: The Joint Staff, September 17, 2006.

Joint Publication 3-33, *Joint Task Force Headquarters,* Washington, D.C.: The Joint Staff, February 16, 2007.

Joint Task Force 536, *Base Order,* Washington, D.C.: The Joint Staff, January 7, 2005.

———, Untitled, brief, Second Fleet, not dated, p. 34.

Joint Task Force–Atlas Response (JTF-AR), *Joint Manning Document,* not dated. Quoted in unpublished research by Armando X. Estrada and Janice H. Lawrence, on Joint Task Force operations, 1990–2004, at the Graduate School of International Business and Public Policy, Monterey, Calif.

Joint Task Force/Combined Support Force–Unified Assistance (JTF/CSF–Unified Assistance), *Joint Manning Document,* December 31, 2004.

Joint Task Force–Noble Anvil (JTF-NA), *Joint Manning Document,* not dated. Document provided by Historian's Office, U.S. European Command, Stuttgart.

Multinational Planning Augmentation Team, "Multinational Force Standing Operating Procedures: Overview Brief," not dated. As of January 1, 2008:
http://www1.apan-info.net/mpat/MPAT/AboutMPAT/tabid/3716/Default.aspx

———, "Multinational Planning Augmentation Team (MPAT): What Is MPAT?" briefing, October 1, 2007a.

———, "Multinational Planning Augmentation Team (MPAT) and Multinational Force Standing Operating Procedures (MNF SOP) Programs," Information Paper, PACOM J722, October 1, 2007b.

"NAVCENT MHQ w/MOC," briefing, March 27, 2007.

"PACOM PLANORD [Planning Order], 271115Z Dec 04, HQ PACOM to COMARFORPAC [commander of the Army Forces in the Pacific]," quoted in Operation Unified Assistance Chronology.

Poynor, Bob, "Is Air Force Command and Control Overly Centralized?" Montgomery, Ala.: Maxwell-Gunter Air Force Base, Air University, June 20, 2007. As of October 14, 2008:
http://www.maxwell.af.mil/au/aunews/archive/0215/Articles/IsAirForceCommandandControlOverlyCentralized.html

School of Advanced Air and Space Studies (SAASS) home page. As of November 13, 2007:
http://www.au.af.mil/au/saass/

USAF Expeditionary Center, "COMAFFOR Senior Staff Course Schedule," Fort Dix, N.J., April 2007.

U.S. Air Force, Global Cyberspace Integration Center, "History of the Global Cyberspace Information Center," Langley Air Force Base, Va., updated January 2008. As of October 14, 2008:
http://www.gcic.af.mil/library/factsheets/factsheet.asp?id=7959

U.S. Army Training and Doctrine Command, *The Army Modular Force,* Fort Leavenworth, Kan.: Combined Arms Center, FMI 3-0.1, November 17, 2006.

U.S. Department of Defense, *Report to Congress: Kosovo/Operation Allied Force After-Action Report,* January 31, 2000, p. 18.

———, *Quadrennial Defense Review Report,* Washington, D.C., September 30, 2001.

———, *Quadrennial Defense Review Report,* Washington, D.C., February 6, 2006a.

———, *Department of Defense Strategic Plan for Joint Officer Management and Joint Professional Military Education,* Washington, D.C., April 3, 2006b.

U.S. European Command, "USEUCOM Training Transformation in Support of C2 Transformation," briefing, September 2003.

———, "The Establishment, Evolution, and Accomplishments of the United States European Command," October 28, 2008 (date last modified). As of November 24, 2008:
http://www.eucom.mil/english/Command/history.asp

U.S. Joint Force Command, *Expanding the Joint C2 Capability of Service Operational Headquarters,* Strategic Planning Guidance 2006–2011 Directed Study Task 04, February 17, 2005.

———, "Standing Joint Force Headquarters (Core Element)," briefing, COL Douglas K. Zimmerman, USA, January 5, 2007.

U.S. Navy, Second Fleet, untitled MHQ with MOC brief, not dated.

———, *Maritime Headquarters with Maritime Operations Center Concept of Operations* (MHQ with MOC CONOPS), Final Draft Version 2.4, Norfolk, Va., October 31, 2006.

U.S. Senate, *Hearing of the Senate Committee on Armed Services: Lessons Learned from Military Operations and Relief Efforts in Kosovo,* Washington, D.C.: U.S. Government Printing Office, October 21, 1999.

Wehrle, Maj Gen Joseph, "Joint Task Force–ATLAS RESPONSE: Commander's Perspective," briefing, not dated.

Welch, CAPT Rodger, USN, "U.S. Military Relief Efforts for Tsunami Victims," U.S. Pacific Command, Camp H. M. Smith, Hawaii, January 5, 2005.